D1029194

About Island Press

Island Press is the only nonprofit organization in the United States whose principal purpose is the publication of books on environmental issues and natural resource management. We provide solutions-oriented information to professionals, public officials, business and community leaders, and concerned citizens who are shaping responses to environmental problems.

In 2002, Island Press celebrates its eighteenth anniversary as the leading provider of timely and practical books that take a multidisciplinary approach to critical environmental concerns. Our growing list of titles reflects our commitment to bringing the best of an expanding body of literature to the environmental community throughout North America and the world.

Support for Island Press is provided by The Bullitt Foundation, The Mary Flagler Cary Charitable Trust, The Nathan Cummings Foundation, Geraldine R. Dodge Foundation, Doris Duke Charitable Foundation, The Charles Engelhard Foundation, The Ford Foundation, The George Gund Foundation, The Vira I. Heinz Endowment, The William and Flora Hewlett Foundation, W. Alton Jones Foundation, The Henry Luce Foundation, The John D. and Catherine T. MacArthur Foundation, The Andrew W. Mellon Foundation, The Charles Stewart Mott Foundation, The Curtis and Edith Munson Foundation, National Fish and Wildlife Foundation, The New-Land Foundation, Oak Foundation, The Overbrook Foundation, The David and Lucile Packard Foundation, The Pew Charitable Trusts, Rockefeller Brothers Fund, The Winslow Foundation, and other generous donors.

Flames
in
Our
Forest

Flames in Our Forest

Disaster or Renewal?

Stephen F. Arno
and Steven Allison-Bunnell

ISLAND PRESS
Washington • Covelo • London

Library of Congress Cataloging-in-Publication Data
Arno, Stephen F.
    Flames in our forest : disaster or renewal? / Stephen F. Arno, Steven Allison-
Bunnell.
    Includes bibliographical references (p.   ).
    ISBN 1-55963-882-6 (cloth : alk. paper) — ISBN 1-55963-883-4 (pbk. :
alk. paper)
    1. Forest fires—Environmental aspects—West (U.S.) 2. Forests and
forestry—Fire management—West (U.S.) 3. Fire ecology—West (U.S.)
I. Allison-Bunnell, Steven. II. Title.
    SD421.32. W47 A76 2002
    577.2—dc21   2002000922

British Cataloguing-in-Publication Data available.

$45.00   1016554

Contents

Preface

Fire has shaped forests in the western United States for thousands of years. As communities of trees, other plants, animals, and microorganisms, these forests are as well adapted to periodic fires as they are to the climate and the native soils. People love these forests, but we have difficulty understanding and accepting the idea that we cannot simply remove fire from them.

If you are interested in western forests, and their future health and prosperity, this book is for you. Our goal is to help you appreciate the importance of fire in these forests and to show how some of the benefits of natural fire can be restored. Restoration involves creative use and suppression of fires that occur, prescribed burning, management of forest fuels, and silviculture—the art and science of tending the forest. We also need to make forest homes and their surroundings more fire-resistant.

This book is a nontechnical account of the fascinating story of fire in our forests that has unfolded from current scientific knowledge and experience. For students and others interested in further information, references to publications that elaborate on specific subjects are cited in the text (in parentheses) by author and date and are listed alphabetically in the References section. Additionally, this book offers candid commentary on wildland fire management practices and policies—arguably one of the most compelling ecological and natural resource management issues facing the West today.

The Forester's Perspective (Stephen Arno)

My early experience with fire was positive, perhaps making it easier to later appreciate fire's role as a vital force of nature. Growing up on the outskirts of the small city of Bremerton on Puget Sound in western Washington, surrounded by humid forests, one of my earliest and most frequent chores was gathering and preparing firewood to help heat our home. I went with friends on frequent camping trips, either by small boat exploring remote areas in Puget Sound or hiking in the nearby Olympic Mountains. The weather was often cold and damp, and we relied on campfires for drying out, keeping warm, and cooking. The wood, like the forest, was usually wet. Thus it was no small feat to build a robust campfire or even an effective blaze in our home fireplace. Forest fires were a possibility, but they were uncommon in those cool, moist woods. To me, fire was a useful tool for keeping warm as well as for disposing of brush and tree limbs. We even used bonfires to crack and break up the giant boulders on our beach that were a hazard for boating.

Leaving home broadened my experience with fire. In 1962 I took a summer job in the woods in a drier part of the state and wound up fighting a few small forest fires, which seemed like an arduous, dirty job, even compared to the outdoor labor I was used to. The next summer, while working in Sequoia and Kings Canyon National Parks in the California Sierra Nevada, I participated in holding the upper fire line on a fast-moving "controlled burn" that raced through several hundred acres of tinder-dry, brushy chaparral on a foothill ranch below the national parks. This was an amazing experience for a youth from the humid Northwest forest. At Sequoia and Kings Canyon I was also introduced to the ecological story that frequent fires had helped perpetuate the magnificent groves of giant sequoias over millennia.

In 1965 my wife, Bonnie, and I moved to western Montana, where I studied forest ecology as a graduate student. At first, I assumed that most fires in Rocky Mountain forests were severe, stand-destroying conflagrations. Big wildfires were a frequent topic of conversation. But my assumptions about fire in the forest were challenged in 1971. My first assignment as a professional forest ecologist involved surveying old growth forests of all kinds, gathering and analyzing evidence of fire history and fire's effects on stand composition and structure. Surprisingly, I discovered many of these forests contained trees that had survived one or more fires in past centuries. The fire-resistant trees—often shielded from damage by thick bark—were of several species. They bore blackened scars at their bases, clear evidence that many fires in the past had been of low to moderate intensity. These fires had *not* killed all the trees in their path like the high-intensity, stand-replacing infernos portrayed in stories and news accounts.

That same year Bonnie and I bought a 40-acre parcel of second-growth ponderosa pine forest land in the Bitterroot Valley of western Montana. Most of the trees on the property had become established soon after heavy logging in the 1880s. The few remaining old growth pines were being logged on the adjacent property. The fresh stumps revealed an unmistakable pattern of frequent low-intensity fires that had left multiple fire scars on these trees, just like the pattern reported in the giant sequoia groves and in ponderosa pine forests elsewhere in the West. Soon I started thinning and using fire to reduce fuels in our family forest, which we eventually expanded to 60 acres. Thankfully, it was relatively easy to keep a fire from traveling out of bounds using standard fire-control tools, provided we burned under appropriate weather conditions, in moderately moist fuels.

I had heard about the first program to allow natural (lightning) fires to burn, which began in 1968 in the high country of Sequoia and Kings Canyon National Parks. This program was a significant accomplishment, but forests of the High Sierra Nevada were generally discontinuous, with intervening barren alpine ridges and rock-walled canyons commonly slowing the growth of fires. It would be a bigger challenge to allow fires to burn in denser forest landscapes.

In the early 1970s some of my Forest Service colleagues were designing a program to allow some lightning fires to burn at all elevations in the Selway–Bitterroot Wilderness of northern Idaho. In 1973, with the program in place, the adjacent White Cap and Fritz Creek fires were allowed to burn amidst great uncertainty as to the outcome, and considerable controversy. By the time autumn rains extinguished them, these fires had covered 2,800 acres, mostly in the warm, dry lower-elevation forests. To my surprise, upon visiting this burn the following year, I found a complex mosaic of different burn severities laid out on the rugged terrain as a result of several weeks of unsuppressed burning.

By the early 1970s the public was becoming increasingly critical of extensive clearcutting in Northern Rocky Mountain forests, and foresters were looking for other ways to harvest trees and regenerate forests for the future. I was on a team assigned to develop advice for managing forests in a manner consistent with the natural processes that had shaped them over past centuries. A few foresters in other parts of the West began to advocate use of selective cutting coupled with prescribed fire in ponderosa pine forests as a means of mimicking the effect of frequent fires of the past. These treatments were designed to perpetuate open-canopied forests containing trees of many sizes and ages while yielding timber for harvest. The concept made sense, and I soon started using variations of it on our family forest as well as working with Forest Service colleagues to test similar practices on national forest land.

Unlike most conventional logging, attention was focused on leaving a stand of large vigorous trees and removing the smaller, slow-growing "suppressed," and diseased trees. We tested low-impact harvesting techniques, such as use of horses and farm tractors equipped for logging, to minimize damage to the soil and the remaining trees. We removed or burned the accumulations of branches in piles and then used low-intensity prescribed fire in the stand to reduce surface fuels and to stimulate tree regeneration and native plants used for wildlife habitat. This was a fire that burned the entire stand with flames mostly 1 to 2 feet high.

The more I used and studied prescribed fire, the better this form of management looked compared to the alternative of excluding fire and then having to cope with the consequences. These consequences include decline of important fire-dependent tree and undergrowth species, increasing density and stagnation of forests, epidemics of insects and diseases, and the high potential for severe wildfires. There are many ways to use prescribed fire and other fuels management treatments to restore more historic conditions or vigorous self-sustaining forests. Our widespread practice of ignoring, and in a sense denying, the original role of fire is costly in terms of decline of ecological values of the forest, including natural biodiversity, as well as economic loss and human suffering—loss of forest homes and property. This book explains these problems and presents viable solutions. Flames in our forest are inevitable. Whether their effects are perceived as disaster or renewal depends on how we manage fire, fuels, and the forest itself.

The Science Writer's Journey (Steven Allison-Bunnell)

In retrospect, I have appreciated stewardship forestry all of my life without knowing what it was called or fully understanding the ecological implications of its practice. Although formal knowledge would only come much later, I absorbed the process and results of stewardship forestry over many visits to my grandparents' woodlot and tree farm in the Willamette Valley during childhood and adolescence. In the late 1960s my father's parents bought 21 acres of what had once been an oak and madrone river bottom forest, but which was by then second-growth Douglas-fir and a field of Christmas trees. By the time I can remember anything about the place, they had been at work for several years pruning dead limbs up every tree trunk, collecting and burning piles of slash and brush, cutting wind-fallen logs and standing snags into firewood, and planting and pruning the Christmas trees. Even though they had a large garden in town to take care of, they spent many happy days working there, sometimes camping in their motor home. For more than 20 years, my grandparents fought off invasions of Himalayan blackberries and Scotch broom and resurgences of ever-present poison oak. It was a family ritual to visit the tree farm at Thanksgiving to cut a Christmas tree, and when I lived with my grandparents during college, I would help my octogenarian grandfather cut firewood (he still handled the chain saw).

The "tree farm" as I knew it was a lovingly tended forested park with widely spaced trees and wild irises, shooting stars, and Calypso orchids growing in the sunny openings beneath the tree canopy. It was home to deer, wild turkeys, and, much to my childish delight, orange-bellied salamanders creeping through the ferns and moss. Although my grandparents were primarily intent on keeping their woods tidy and bringing in as much firewood as they needed to heat their home during the damp Oregon winters, they also created a wonderfully diverse and attractive habitat. Was it natural? Certainly they managed it quite intensively. But save for a few rotting stumps and narrow tracks winding through the woods, it looked quite unharmed. Any doubts as to the value of their active management were dispelled when they could no longer maintain the Christmas tree stand and the small Douglas-firs grew to 20, then 40 feet tall. Ten years after the last Christmas tree was cut, a dense thicket of firs covered the area so tightly that the only thing growing on the ground was moss and a few bracken ferns. Now that both of my grandparents are dead, the tree farm has passed into my father's keeping. Although he can't spend the same amount of time on the place as they did, he has arranged for this area to be thinned. Now that I know what I do about forest ecology, we look forward to seeing how this young forest matures alongside its elder counterpart in the woodlot.

Broadcast burning or prescribed fire was not among the tools my grandparents used in their woodlot. My grandmother took much pleasure in burning the brush and slash in huge bonfires, but she always burned in the same place and only when it was safe to do so. I now know that what she did was a classic fuel reduction treatment in a fire regime that would probably have burned in a high-intensity stand-replacement fire if left alone. None of us had ever heard anything besides the idea that forest fires were dangerous and undesirable. The comic book version of the Smokey Bear saga was distributed in my school and I treasured my copy. In my mind's eye, my father's stories about his stints on summer fire crews in the Oregon Cascades during his college years were surrounded in an aura of heroism fighting the good fight even though he actually spent more time digging hiking trails than fireline.

At the same time that I saw what the involved and patient owners of a small woodlot could accomplish on the scale of an individual stand, I grew up with a profound dislike for the results of industrial logging as it was practiced in the Coast and Cascade Ranges of Oregon and Washington. Whereas my grandparents had minimized their impact, in commercial clearcuts I saw hundreds of acres of stumps, bulldozer-scarred hills, and tangles of wasted timber and slash. It was a sobering moment when on a family vacation, we counted the rings on an enormous section of a log left behind and realized that the tree had been growing before

Columbus set sail. I knew that I would never see the likes of that tree again in my lifetime. Ken Kesey's description of the beauty and thrill of felling trees in *Sometimes a Great Notion* is the only depiction of logging that could convince me it was anything other than a soul-numbing brute rape of the land. Robert Leo Heilman's memoir of a tree planter, *Overstory Zero*, only served to reinforce my previous perception.

I remained an unapologetic tree-hugger and unquestioning fire-hater until I moved to the Northern Rockies and met Steve Arno while reporting a series of articles on wildland fire ecology for the Discovery Channel Online in 1997. Like most writers unaware of the details, I was initially attracted to the story because of the exploits of the smokejumpers conveniently based in Missoula, and because my academic training in environmental history and the social studies of science conditioned me to ask how we know what we know about that natural world. When even the smokejumpers—the people most highly trained to fight fires—told me that they would rather be involved in prescribed fire and restoration treatments than suppression, I knew this dreaded enemy deserved closer attention. Interviewing Steve and other advocates of reform in fire management policy like Ron Wakimoto of the University of Montana School of Forestry, I was astonished to have all of my assumptions about both fire and logging turned inside out. I tried to convey that sense of a personal paradigm shift in my original articles and am grateful for the chance to amplify it further in the much finer detail that this book offers.

Today I strongly believe that fire and logging do not inherently contradict environmentalist values. In one of my first conversations with Steve Arno, he said that we must become "high-tech hunter–gatherers" as we manage and adapt to the forests of the West. That phrase has stuck with me as an encapsulation of the idea that we can both use and respect the land. I believe this is possible because I no longer believe in the concept of untrammeled wilderness completely devoid of human presence or impact. Recent scholarship in environmental history has shown that human beings have been interacting with and shaping the landscapes they inhabit for as long as we have evidence to tell the story. Our modern destruction of habitat and natural places differs only in scale and intensity. The American conservation movement was founded on the myth of wilderness as uninhabited, pristine nature contrasted sharply with the decadence and unnaturally stultified life of cities. We now acknowledge what Progressive Era environmentalists did not: that American Indians shaped the landscapes whites saw as untouched by human hands, and that fire was their primary tool. So there is no such thing as pure nature, no human-free wilderness, only lands where we work in concert with, instead of in ignorance of, natural processes. Many committed environmentalists reject this idea as the first push on a fast ride down the slippery slope of business as usual exploitation of the land without regard for biological diversity, ecosystem health, or the intangible but profound spiritual value of wildness and wild places. We do not presume to convince the most adamant, but we do hope to make it clear that none of what we propose is business as usual in approach, motivation, practice, or outcome. My own journey to this perspective has been too full of surprises for that to be the case. Instead, this book is a manual for would-be high-tech hunter–gatherers. Steve Arno's voice and expertise speak authoritatively throughout this book, and it is a privilege to have been his collaborator. We warmly dedicate it to all stewards of our forests, past, present, and future.

Acknowledgments

We wish to thank reviewers who read early drafts of this book and provided suggestions leading to substantial improvements: Jim Agee (University of Washington), Matt Arno (forest restoration contractor), Steve Barrett (fire ecology consultant), James K. Brown (Intermountain Research Station, retired), Carl Fiedler and James R. Habeck (University of Montana), Mick Harrington and Jane Kapler Smith (Rocky Mountain Research Station), Bruce Kilgore (National Park Service, retired), Dennis Knight (University of Wyoming, retired), Bob McKee and Brooke Thompson (Bitterroot National Forest, retired), Dave Parsons (Aldo Leopold Wilderness Research Institute), Carl Skinner and Phil Weatherspoon (Pacific Southwest Research Station), Cathy Stewart (Lolo National Forest), Janie Canton-Thompson (environmental writing contractor), and Diana Tomback (University of Colorado, Denver). We thank Jack Losensky, Hamilton, MT, for lending us historic photographs, and Sara Arno for editing and typing the reference section. We also greatly appreciate Stephen J. Pyne's monumental contribution to the history of forest fire in American culture.

Chapter 1

Introduction:
Why Learn about Fire?

Since the end of the last ice age fire has molded most forests in the western United States. Repeated patterns of burning heavily influenced which tree and undergrowth plant species prospered. Fires also helped determine the structures and patterns of forests on the landscape as well as their suitability as habitat for a myriad of birds and mammals. Pioneer western naturalists were sometimes alarmed by the destructive power of fire; but they also recognized that fires helped create many of the splendid old growth forests at which they marveled (Pinchot 1899). Now, for almost a century, we have tried to eliminate fire from most forests without considering possible adverse effects on native plants and animals and on the sustainability of the forest itself. Logging and other human activities have of course changed western forests dramatically. But our endeavors to eliminate fire have caused some of the most widespread and harmful changes.

This book explores the underlying historical and ecological reasons for the problems associated with fire exclusion. It also examines the scientific knowledge and available technology that could be used to mimic and restore some of the ecological effects of natural fires to help sustain western forests. The messages about forest fires that we have been exposed to for nearly a hundred years have bred skepticism as to the possibility, or even the desirability, of restoring some approximation of the natural fire

1

process in western forests. This book presents a new picture of fire's role in maintaining forests, what options we have for restoring important effects of historical fires, and the consequences of not doing so.

To understand why fire is so universally feared, we begin with the history of European-American perceptions and uses of fire in the forest. (It may be surprising to learn that a century ago some influential people thought it was actually a good idea to burn the forest regularly.) Until the early twentieth century, the western states and their residents commonly encountered large, free-ranging fires. By midcentury, fire suppression had become quite effective, and the scale of burning had been greatly reduced. However, since the 1970s the extent and severity of fires has been on the rise. We will show how fire suppression has changed the extent of burning on the western landscape, and we will project future trends for fire.

How do forest fires burn? The principal factors and how they influence fire behavior may seem relatively simple, but nature has many ways of confounding our predictions of what fires will do. Different plants and animals have adapted in diverse ways to different kinds of burning. Through natural selection, contrasting patterns of burning were linked, and often fine-tuned, to different types of forests. Some trees produce so much fine fuel they virtually assure frequent burning, which in turn benefits their survival and continued abundance. Other trees are adapted to severe but infrequent fires. Although these fires destroy them, they can regenerate and grow afterward much more effectively than their competitors. Individual trees may not survive the fire, but the species and its unique forest community of associated plants and animals thrives.

Fire is not only a biological phenomenon. Fires at any given location might produce vastly different effects on the soil, water, and air. Depending upon weather and the quantity and moisture of fuels, fire in a given forest may lead to negligible or massive erosion. Heavy erosion after fire is a natural phenomenon in some forests, but in other forests such erosion is related to human alterations. Even fires that have little impact on the natural system can hurt humans, as in the case of smoke pollution. However, through management of vegetation and use of prescribed burning, we can reduce the negative impacts of fire. To help you consider how a forest—perhaps your own local forest—could be protected or restored, we present a method for uncovering the history and effects of past fires in any particular forest.

People and forests can benefit from planned and prioritized prescribed fire and fuel treatments for different zones on the landscape. Intensive fuel treatments are suitable for the residential forest zone; we recommend broader-scale fuel treatments for the general forest zone; and the use or simulation of natural fires is appropriate for large, remote natural areas. Each alternative for managing wildland forests, including continued exclusion of fire, use of natural fires, thinning and fuel removal, and pre-scribed burning, has its own benefits and drawbacks. For people living in the woods, reducing fuels around forest homes and managing the home-site's forest is more important than is often realized. Finally, with all of these historical, ecological, and new land management challenges in mind, we believe that the time is ripe for changing our perception and management of fire-dependent forests. The opportunity to act lies before us but is likely to disappear if we do not seize it soon.

Wildland Forests and Their Fires

Western forests and woodlands cover hundreds of millions of acres of diverse and rugged country in a confusing array of tree species, each adapted to different growing conditions and different frequencies and intensities of fire. Some types of forests burned naturally in frequent, low-intensity fires, whereas others burned infrequently in high-intensity, stand-replacement fires, and still others experienced a range of burning intensities. Still more confounding is that, depending on the type of for-est, a high-intensity conflagration may be a natural phenomenon or the result of human-caused changes in the forest.

The native tree species, undergrowth plants, and animals constitute forest communities that in many cases are direct descendants of forests that developed over at least the past 11,000 years following the last ice age. The assemblages of trees, undergrowth plants, and animals are still primarily influenced by the same processes that shaped them over thou-sands of years. They are called wildland forests to distinguish them from the homogeneous, cultivated plantation forests that are prevalent today in much of the world. Roads, logging, livestock grazing, and residential subdivisions have affected many of our western wildland forests. Still, they are predominantly natural communities.

Westerners interested in the outdoors cannot escape the topic of fire in the forest. Consider our most extensive and heavily populated type of forest. Dominated by the drought-tolerant ponderosa pine *(Pinus ponderosa)*, it stretches from the Great Plains of Nebraska and South Dakota almost to the Pacific Ocean and from northern Mexico to southern British Columbia (Fig. 1.1). Today, hundreds of thousands of people live in the shade of stately ponderosa pines. Historically, in the majority of these areas, frequent low-intensity fires maintained parklike stands of trees mostly pruned of low branches with open grassy understories. Fallen pine needles and cones make an ideal fuel bed capable of burning

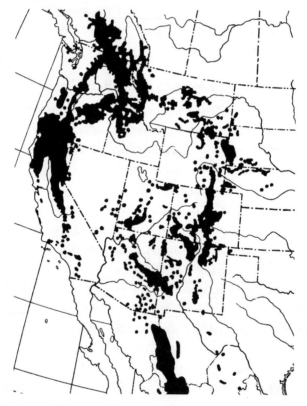

Figure 1.1. The distribution of ponderosa pine forests across the western United States. (Map based on Little 1971.)

throughout much of the year. With protection from fires, needle litter and thickets of small trees have transformed much of this forest into a dense fuel type that burns in severe fires. For example, well after the traditional summer fire season in mid-October 1991, a windstorm-driven fire destroyed over 100 homes in the suburbs of Spokane, Washington. Yet, with good management of forest fuels, ponderosa pine forests and the homes within them can be relatively safe from fire damage.

Anyone following the news no doubt realizes that western forests are prone to burning from time to time. What's usually missing from the stories is the fact that the magnificent old growth forests extolled by naturalists and featured in western national parks were shaped by recurrent fires over thousands of years. Both the giant sequoia *(Sequoiadendron giganteum)* groves high in California's Sierra Nevada and the towering redwood *(Sequoia sempervirens)* forests of the fog-bound coastal lowlands often had open understories as a result of frequent low-intensity fires (Brown and Swetnam 1994, Fritz 1931, Swetnam 1993). Giant sequoias and redwoods have thick-barked trunks over 10 feet in diameter splotched with char from historical fires. Although they are set aside within national parks and other natural areas, it is often difficult to gain public support to allow prescribed fire for maintaining these ancient forests.

Farther north, in the western portions of Oregon, Washington, and British Columbia, much of the original forest was dominated by gigantic coastal Douglas-firs *(Pseudotsuga menziesii* variety *menziesii)*, which were dependent on infrequent fires burning at higher intensities that give this species a competitive advantage over the shade-tolerant ("shade-loving") western hemlock *(Tsuga heterophylla)*. Large areas of old coastal Douglas-fir are protected in national parks and national forests. But by keeping fire out, we are suppressing the primary natural process that perpetuates Douglas-fir.

Inland, in the Rockies and other high mountain ranges, most of the forest trees are adapted in different ways to fire. Some forests contained trees that commonly survived fires. Stand-replacement fires typically killed trees in lodgepole pine *(Pinus contorta)* forests and in the towering stands of Engelmann spruce *(Picea engelmannii)* at higher elevations.

Then trees regenerated from seeds that survived in closed cones or light seeds that blew in from adjacent forests. With protection from fire, these forests often become virtual monocultures of the most shade-tolerant tree accompanied by sparse undergrowth and scant forage for wildlife. In contrast, the historical forests subjected to natural fires often had more diverse mixtures of trees, shrubs, and herbaceous species that could support a richer community of birds and mammals.

Eliminating Fire

Political and social pressures propelled the campaign to eliminate fire in the early twentieth century. To people frightened by the power and destructiveness of wildfires, it seemed logical and necessary to try to remove fire from the forest. However, proponents of the crusade against fire did not understand the long-term consequences of trying to eradicate an intrinsic part of the forest ecosystem. Although natural resource management today strives to protect and preserve wildland forests on public lands, many biologists and foresters have concluded that the forests are instead deteriorating (General Accounting Office [GAO] 1999, O'Laughlin and others 1993). A root cause is that for nearly a century, our policy has been largely based on eliminating fire. Paradoxically, most of these ecosystems depend on characteristic patterns of fires to shape and rejuvenate them. In some of these forests, attempts to remove fire have led to overly dense, stagnant tree growth, which ultimately leads to fires of extreme severity that are dangerous and costly, if not impossible, to control (Ferry and others 1995, Williams 1995).

Ecological science makes it clear that fire is just as integral to the life of western forests as wind, rain, and sunshine. But, unlike other forces of nature, fire's role in the forest can be controlled or altered by humans. Are we employing our best knowledge as the basis for managing fire in the forest? No, according to an analysis by Congress's General Accounting Office (GAO 1999). This is not a new revelation. In the early 1900s several private timberland owners argued that practical forestry should employ fire as a management tool (Hoxie 1910), and a pioneer in the newly emerging science of forest ecology urged the Forest Service to use controlled burning in management of lodgepole pine forests (Clements 1910).

Our counterproductive relationship with fire became manifestly obvious during the record-setting wildfire season in the summer of 2000. Federal agencies spent about $1.6 billion combating fires in the western states (Perkins 2001). That amount does not include substantial costs to rehabilitate bulldozed firelines, plant trees, and prevent or repair runoff damage related to fire. In spite of that effort, over 5 million acres burned. Thousands of dollars per acre were spent fighting fires that threatened residential areas in the forest. Moreover, high-intensity burning and bulldozed fire-control lines damaged soils. In many of these forests, fire history studies suggest that under natural conditions, fire damage to soils was minor, and many trees typically survived fires (DeLuca 2000).

Understanding Fire in the Forest

Because of their close relationships with fires, western forest ecosystems are considered fire dependent. If we hope to sustain the communities of trees, plants, and animals that characterize these wildland forests, we need to understand the natural role of fire, changes brought about by suppressing fire, and alternatives for restoring some reasonable semblance of the natural fire process.

Understanding the role of fire is a mind-stretching exercise not only because it embraces the concept of working with nature instead of controlling it, but also because it relies on highly technical scientific information. Managing wildland forests with fire is an ancient technique used by subsistence farmers and hunter–gatherers worldwide (Pyne 1997). These people, who absolutely depended on nature, employed fire for a variety of purposes, including increasing the abundance and vigor of plants used for food and forage (Lewis 1985, Pyne 1982). At the same time, the ecological effects of fire are so complex and intricate that even the most sophisticated computer models provide only rough approximations (Keane, Ryan, and Running 1996).

Those who revere wildland forests desire to protect them from all human influence. In recent history, preservation has usually included fire suppression, which, as already described, is a radical departure from the natural processes that produced the forests. Given this irony, should we just allow natural fires to return? This can work to some extent in the largest and most remote natural areas. However, many factors make this proposition extremely costly and dangerous for most forestlands. First, fires historically spread across large landscapes by way of broad valleys covered with dry grass and other combustible vegetation. Today these valleys are no longer available for burning due to grazing and other agricultural uses, roads, suburban homes, and other development. Second, many wildland forests at lower or middle elevations have now missed three or more natural fires as a result of fire suppression, and many have also been altered by logging. As a result, these forests differ so greatly from historical forests in both stand age structure and species composition that now lightning fires often burn more intensely than fires of the past. Finally, people have built hundreds of thousands of homes in and adjacent to wildland forests in the past 30 years. Few of these homes use fire-resistant building materials, and few homeowners have reduced fuels in the surrounding forest. To live safely in these conditions, we have almost universally attempted to eliminate fire rather than adapt to a fire-prone environment. (Those who belong to the small minority living in a fire-resistant house in a defensible area deserve congratulations.)

Restoring Fire

One potential solution to our dilemma with fire in western forests lies in acknowledging fire's important ecological role and directing our efforts toward restoring some representation of natural fire (Arno and Brown 1989, GAO 1999). This would involve many kinds of treatments designed to meet restoration needs in different forests, some of which were historically characterized by low-intensity fires and others by high-intensity conflagrations. In some locations remote from homes and private lands, treatments could include managing lightning fires rather than suppressing them. Treatments can be designed to restore fire's natural influence to help maintain forests that are sustainable through time for a variety of purposes, including habitat for wildlife, watershed protection, and reduced hazard of severe fires to developed areas. Many foresters, ecologists, government leaders, and community groups support this restoration concept, which both Congress and the White House endorsed in the fall of 2000. Barriers to this undertaking are huge, but not insurmountable. They include finding ways to finance restoration and dealing with environmental and smoke regulations that complicate prescribed burning. For the past hundred years, forest management regulations and procedures have been based on the assumption that fire is a damaging agent rather than a vital force of nature. The institutional culture of the agencies responsible for stewardship of wildlands emphasizes fire suppression over management of fire and fuels. Nevertheless, the major federal land management agencies in both the United States and Canada have the knowledge and technical ability to restore fire in some beneficial form. Implementing any appreciable amount of this forest restoration will require a high level of public support.

Many people with long experience in the woods see such restoration as a win–win scenario that could provide continuing benefits such as improved habitat for wildlife and recreational opportunities, and products such as clean water and other ecosystem services, fish and game, lumber and fuelwood. In contrast, others argue that "restoration" is a subterfuge to reopen federal lands to road building and massive commercial logging. We agree that these concerns must be addressed by ensuring that restoration is guided by broadly agreed upon goals, based on state of the art ecological knowledge and supported by good administration and oversight.

Americans cherish the vast wildlands found in national forests, national parks, and other large reserves in western North America. Publicly supported legislation and regulations mandate that most federally owned forestland be kept largely undeveloped as wildland and managed to perpetuate native plants and animals living in natural ecosystems (Fig. 1.2). Wildland forests of the western United States and similar reserves in Canada are maintained in public trust for present and future generations at considerable expense. Some people view management of the vast area of government-owned land in the western United States as an infringement on their individual freedoms and property rights. Because of the public funding and controversy associated with these forests, establishing effective management is both important and challenging. Demonstration of good management of these wildland forests is necessary for validating the claim that government control of these lands is worthwhile and not merely coercive.

Across a spectrum of issues and concerns, the largest challenge facing

Figure 1.2. The black-backed woodpecker, listed as a sensitive species by the Forest Service, seeks out the insects colonizing dead trees in recently burned forests. (Milo Burcham photo.)

government land management agencies and even private owners who want to manage their forest in the new era of forest restoration may not be expertise or even funding, but credibility. In this book, we cannot claim to create instant credibility for forest restoration, but we hope to highlight viable alternatives that make it plausible.

Chapter 2

Mixed Signals:
A Brief History of American
Perceptions of Fire

The prominent role of fire in mythologies of cultures around the world reflects a shared regard of fire as both an essential friend and a destructive foe (Pyne 1982, 1997). For over 50 years, Smokey Bear and Bambi have been a modern mythology familiar to most Americans that casts fire in an entirely negative light. Smokey sternly warns, "Only you can prevent forest fires" (Pyne 1982). Smokey's messages are often delivered superimposed over the graphic scene of a forest devastated by wildfire. This fabulously successful advertising campaign has rejected any suggestion that fire is a necessary component of natural forests.

In 1944, one year before the Smokey character was created, the U.S. government's Wartime Council used Walt Disney's Bambi in the national fire prevention campaign. In the movie *Bambi*, released in 1942, one of the most frightening scenes shows the heroic deer trying to escape a forest fire caused by a "hunter" portrayed as a wicked poacher (Lutts 1992). Horror writer Stephen King says that the Disney cartoons, including *Bambi*, frightened him most as a child (Lutts 1992). These powerful icons give form to our fear of fire without any countervailing imagery illustrating its value.

From the time the first European-American settlers arrived in the American West, they struggled with forest fires. Civilization seemed pitted against untamed nature in a vast region of wild forests, rugged mountains, semiarid climate, and inevitable conflagrations. Ever since the limitless western frontier passed into history in about 1890, many conservationists, politicians, and other prominent citizens have advocated eliminating fire as a threat to society and forests. Conversely, some woodsmen and forest landowners argued that fire is an inherent part of the wildland environment that should be used to human advantage. Changes in philosophy about fire in the forest define a series of relatively discrete historical eras.

Accepting Risk and Adapting to Fire: Ancient Times to 1910

At the beginning of the twentieth century, when most of its states and cities had been established, the American West was still a place where fire roamed free. Pioneers in the West, like early peoples worldwide who made their living off the land, had a pragmatic view of fire. Although they may not have shared the mysticism of native cultures, settlers were compelled to acknowledge fire's power and importance. From the dawn of human history through the early twentieth century, people worldwide used fire in agriculture and to help clear forestland. They burned wood to heat their dwellings and cook their food. Still, humans have probably always had a love–fear relationship with fire in the forest as evidenced by the importance of fire in the legends of ancient Greek and American Indian cultures (Pyne 1997). People today, no matter how technologically advanced, still regard fire in the forest with awe.

Preindustrial peoples worldwide used fire in the forest as their one and only powerful tool (such as slash and burn agriculture) to mold the environment to their advantage (Pyne 1997). During the Renaissance, farmers in the mountains north of Venice, Italy, commonly set fire to the forest to clear more land for grazing and crops (Perlin 1991). The forests associated with ancient civilizations in the Middle East and Europe were generally dominated by fire-dependent tree species such as oak and pine. Pollen and charcoal sediment layers in lakes and soils provide evidence of periods 4,000 to 5,700 years ago when people across Europe set fires for agricultural purposes (Carcaillet 1998, Clark, Merkt, and Muller 1989, Tinner and others 1999).

In the first half of the nineteenth century, American Indians as well as the new arrivals to the western frontier—trappers, adventurers, and pioneers of all sorts—used campfires every day, sometimes carrying hot coals with them as they traveled. Some of these campfires no doubt developed into forest fires. American Indians also purposefully set fires to maintain valuable features of forests in western North America. Near the Pacific coast from southern British Columbia to central California, native peoples burned the small grassy areas surrounded by forest that early European-Americans referred to as prairies. This attracted deer for hunting and perpetuated a variety of important food plants (Boyd 1999). Farther inland, some tribes set large, uncontrolled fires for many purposes, including propagating food or forage plants, driving game into an ambush, clearing travel routes and campsites, signaling a gathering or council, and driving enemies out of dense hiding cover (Arno, Smith, and Krebs 1997, Barrett and Arno 1982, Gruell 1985a, 1985b, Lewis 1985). Early European-Americans in the West often used fire to help clear forest homesteads and to improve livestock grazing in the forest. Prospectors sometimes set the mountain forests ablaze to expose mineral deposits.

As the West became increasingly populated with towns, mining camps, and other settlements, forest fires were increasingly viewed as a threat. Virtually all buildings in forested regions were made of wood, often including wood shingle roofs, and were highly combustible. Frequently they were set afire by sparks from a woodstove or from other accidents. Once a fire started in wooden settlements, it could easily engulf most of the community and, in the dry season, burn far and wide through adjacent forests. Railroads were notorious for igniting forest fires with the sparks that spewed from locomotive smokestacks and clanking iron wheels (Pyne 2001a). Logging often left huge accumulations of dead, dry limbs and tree crowns surrounding forest settlements, which allowed conflagrations to develop rapidly in hot, dry weather. In 1889 fires burned millions of acres of forest in the inland West as well as major portions of Seattle, Spokane, Ellensburg, and parts of several other northwestern communities (Barrett, Arno, and Menakis 1997, Taylor 1989).

There were no effective means of suppressing forest fires until well after 1910, and people in the West were all too familiar with the catastrophic fires that had swept the North Woods claiming hundreds of lives. On the same day as the Great Chicago Fire, October 8, 1871, forest fires destroyed Peshtigo and smaller villages in Wisconsin, killing 1,500 people and charring 3.8 million acres (Davis 1959). Ten years later, forest fires in Michigan killed 169, and in 1894 the Hinckley Fire in Minnesota produced a death toll of 418.

Forest fires in the American West also threatened human life. In 1902 the Yacoult and nearby fires in the western portions of Washington and Oregon burned over a million acres and claimed 38 lives (Davis 1959, Pyne 2001a). But the environment and fuels around many settlements in the interior of the West were different from those in the humid region west of the Cascades and the country east of the Mississippi. Much of the interior West is a semiarid landscape that historically supported grassland, sagebrush, and other desert shrubs or relatively open forests of ponderosa pine and other trees. Here, dry grass allowed fires to spread, but early in the settlement era, unrestricted livestock grazing removed much of the grass. Cultivation and flood irrigation soon occupied much of the valley lands, further restricting the spread of fires. In many areas American Indians continued their burning practices during the late nineteenth century (Barrett 1981, Evans 1990). As a result, fuels in many western forests and adjacent grasslands remained relatively sparse into the early twentieth century. Some local people recognized that in some of the western forests and grasslands there were opportunities to use fire to control fuels as well as for other benefits.

However, with increasing development, exploitation of forests, and the rise of the conservation movement, a different view prevailed. Forest conservation became a major Progressive Era cause at the turn of the twentieth century, and growing public concern about the welfare of forests contributed to the crusade against fire. The conservation movement was largely a reaction to the often profligate exploitation of forests in the East and Midwest (C. Miller 2000). In the mid nineteenth century, American lumberjacks cut more than 100 million cords of wood (a cord being approximately a ton of dry wood) each year to fuel smelters, steamboats, and railroads and to heat homes across the country (Perlin 1991). Large areas of forest were also cut for lumber, the most common building material for the rapidly growing nation. This massive wood use removed a large percentage of the forest east of the Great Plains and fueled fears of a "timber famine" (C. Miller 2000). Burning often followed logging as settlers sought to clear land for pasture or plowing. By the late nineteenth century, Americans had seen magnificent forests of the East, South, and Midwest destroyed in the name of profit and progress. It was clear this fate also awaited western forests unless the federal government could be persuaded to stop it. This wholesale exploitation of forests became the impetus for the creation of over 200 million acres of federal forest reserves (later renamed national forests) and national parks starting in the 1890s.

Although timber harvesting was arrested on the new forest reserves, destruction of forests by fire was not. Compared to other natural hazards of the time such as drought, severe winter storms, hurricanes, floods, or untreatable diseases, danger from forest fire was not noteworthy or special. However, those in the new profession of forestry as well as some political leaders held a different view. One early western forester lamented the common attitude of local people: "Fires were treated with no concern, and men failed or refused to assist in their suppression" (Shoemaker 1922).

Forestry developed as a profession in northern Europe during the nineteenth century. Its original goal was to grow trees on abused lands where the native forest previously had been denuded by exploitation for timber, fuelwood, or clearing for pasture. The climate was humid, and fire was not known to be an important natural factor in forest development. Instead, European foresters viewed fire in the forest as unnecessary and destructive. This attitude came to the United States in the late nineteenth century along with other European concepts of the newly developed profession of forestry. Gifford Pinchot, a young man from Pennsylvania whose grandfather had been a timber baron there, studied forestry in Europe and brought it to the United States (Pinchot 1947).

While trying to establish forestry to conserve American forests, Pinchot traveled widely in the western United States. He observed that fire had played an important role in producing some of the magnificent natural forests. In 1899 he published an essay, The Relation of Forests and Forest Fires, in *National Geographic* magazine. He urged that the role of fire in creating forests be studied to help in designing forest management (Pinchot 1899). Pinchot wanted to prevent destructive fires, but he also wanted to understand fire's role as an ecological force. He remarked, "A few observers who have lived much with the forest, such as John Muir of California, have grouped fire with temperature and moisture as one of the great factors which govern the distribution and character of forest growth" (Pinchot 1899, 393). Pinchot indirectly recognized the ecological value of fire when he observed that if fires had been kept out of the majestic coastal Douglas-fir forests of the Northwest, they "would be replaced in all probability by the hemlock, which fills even the densest of the Puget Sound forests with its innumerable seedlings" (Pinchot 1899, 403). At the same time, Pinchot staunchly advocated fire suppression, and the political drive to establish complete suppression of fire ultimately overshadowed the scientific need to study fire's natural role in the forest (Pyne 2001a). Ironically, if the ecological studies that Pinchot called for in 1899 had been conducted, they might well have suggested that fire could be useful in forest management.

In the minds of Pinchot and his colleagues, the immediate need was to protect forests from the menace of unwanted fires, many of which were started accidentally or carelessly by humans (Pyne 1982, 2001a). In 1898 Secretary of Agriculture James Wilson appointed Pinchot to take over the fledgling federal forestry program. When Theodore Roosevelt became president in 1901, his close friend Gifford Pinchot gained influence within the federal government. Pinchot's Division of Forestry became the Bureau of Forestry in 1901. In 1905 it became the U.S. Forest Service and was given the responsibility of protecting the vast forest reserves. As it grew from a staff of only 11 in 1898 to a sizeable agency of 821 employees in 1905, the Forest Service attempted to define its mission and secure long-term funding from Congress.

Suppression versus Light Burning: 1910 to 1935

The U.S. forest reserves had been established to protect watersheds and timber resources from rapacious, unregulated logging. However, fires caused deliberately and accidentally by the influx of people in the West remained a serious threat to both watersheds and timber in the reserves.

The Forest Service was charged with prevention and control of fires, and in 1908 Congress set up a unique system, like an open checkbook, to assure payment for fire suppression efforts as needed (Pyne 1982). Within a few years the Forest Service was using "studies" of fire to provide data for propaganda in its efforts to build a fire suppression program (Pyne 2001a). Contrary results were ignored, such as when pioneer ecologist Frederic Clements called for using fire as a tool in managing lodgepole pine forests (Clements 1910). It would not be until the 1940s and 1950s that a few foresters began studying fire ecology in western forests.

The Forest Service was confident it could control wildfires, but this self-assurance was short-lived. In August 1910, despite the young agency's best organized efforts at fire suppression, hurricane-force winds fanned hundreds of lightning fires into gigantic conflagrations that engulfed forests in northern Idaho and northwestern Montana. Most of the 3 million acres were incinerated in a two-day firestorm that killed 85 people and destroyed one-third of the mining city of Wallace, Idaho, and the small communities of Taft, Haugan, Henderson, and DeBorgia in Montana (Cohen and Miller 1978, Koch 1942, Pyne 2001a).

Much of the forest that burned was located in a moist region of northern Idaho and northwestern Montana where natural fires in prior centuries had occurred at intervals of 100 years or more and had typically burned at high intensity. Where the 1910 fires extended eastward into drier forests dominated by ponderosa pine, they burned at lower intensity like the frequent surface fires of the pre-1900 era. However, subtleties in the different role fire plays in different forests were not then and are still not widely appreciated.

The 1910 fires added flash fuel to an already contentious national debate over forest preservation and fire suppression (Pyne 1982, 2001a). The political uproar began to build one year earlier, when William Howard Taft replaced Theodore Roosevelt as president. While Roosevelt was an ardent conservationist and staunch supporter of Pinchot and the Forest Service, Taft was indifferent to Roosevelt's conservation agenda and Pinchot's forestry concerns. The volatile Pinchot soon engaged in a bitter public dispute with Taft's interior secretary, Richard Ballinger; Pinchot accused him of virtually giving away publicly owned coal reserves to his industrialist friends (Pinchot 1947). Taft fired Pinchot in January 1910, sending political shock waves across the nation. However, the Forest Service's new chief, Henry Graves, and most of its staff remained intensely loyal to Pinchot and carried on his policies when the West burned later that summer.

Not everyone saw fire as the enemy the Forest Service did. In August 1910 *Sunset* magazine published an article by northern California timbermen who advocated for "light burning" to reduce the threat of wildfires (Hoxie 1910, Pyne 2001a). Light burning involved informally setting fire to the forest floor litter across large areas in a "safe" season. This was the forerunner to "controlled burning," with firelines and formal procedures for fire control. In recent decades, "prescribed burning" has become the common practice of land management agencies and involves scientifically based application of fire to accomplish specific objectives for fuel reduction and management of vegetation. Other timbermen had also championed light burning, and Interior Secretary Ballinger supported the idea, stating that "we may find it necessary to revert to the old Indian method of burning over the forests annually at a seasonable period" (Pyne 1982, pp. 262–263). The timing of this article could not have been worse. The same summer, one of the California light burns got out of control and burned 33,000 acres before it was finally stopped at the boundary of a national forest. Also, the disastrous 1910 conflagration in Idaho and Montana made a mockery of the Forest Service's assertion that it could control wildfires. The federal government spent the then unprecedented sum of over a million dollars attempting to control the 1910 fires, but much of the workforce, gleaned from streets and bars in nearby towns, was ill-suited to the arduous work of building firelines (Pyne 1982, 2001a).

Heated controversy in the aftermath of these events gave birth to the Forest Service's resolve to anchor its mission of forest protection to the complete suppression of forest fires. In October 1910, Henry Graves visited the extensive ponderosa pine– and sugar pine–dominated timberlands owned by T. B. Walker in the mountains northeast of Sacramento (Pyne 2001a). Walker's crew methodically underburned large tracts of forest after the first fall rains to reduce hazardous fuels and brush encroachment. Although he couldn't deny the effectiveness of this treatment, Graves didn't like the precedent of people using fire. The fact that Ballinger, Pinchot's nemesis, had supported the practice of light burning helped cement the Forest Service's adamant opposition to it (Pyne 1982).

The events of 1910 motivated the Forest Service's attempt to eliminate fire from the forest, since referred to as the "fire exclusion policy." However, debate about the possible merits of light burning intensified and attracted public attention through the early 1920s, especially in California (Biswell 1989, Pyne 1982). The Forest Service regarded this controversy as a serious threat to its effort to establish forestry in the United States. Light burning was unacceptable to the Forest Service because it was too difficult for its advocates to apply with any consistency. There was no formal knowledge of the interrelationships among fuels, weather, and fire behavior, and foresters trained on the European model saw no value in such studies. Also, the Forest Service argued that light burning killed seedlings and small trees, and therefore had to be detrimental to growing new timber. After 1910, the Forest Service's fire exclusion policy became the law of the land based on the premise that fire was unnecessary and destructive. For several decades the Forest Service fought any suggestion that fire might ever be beneficial (Biswell 1989, Pyne 1982, Schiff 1962).

The War on Fire and the
Rise of Prescribed Burning: 1935 to 1978

From 1935 until the 1970s federal forestry agencies expanded their efforts to conquer fire with fire-fighting technology and public advertising campaigns (Pyne 1982). The stage for a coordinated call to arms against fire had been set earlier by the Weeks Act of 1911, which allowed cooperative agreements and matching funds between the Forest Service and state foresters to broaden fire protection on public and private lands. Then the Clarke–McNary Act of 1924 greatly expanded the cooperative fire protection program. It provided major funding to the states for fire protection, with oversight by the U.S. Forest Service. Clarke–McNary also financed a 1940s mobilization effort that took advantage of World War II technology and surplus equipment to greatly reduce the total area burned annually in wildfires.

Elers Koch was a prominent Montana-born forester who had joined Gifford Pinchot's Bureau of Forestry in 1902. In a 1934 essay, Koch poignantly lamented the effects of fire suppression in the Selway and Lochsa River wilderness of northern Idaho. As a result of his long experience, he concluded that fire suppression in this rugged, heavily forested backcountry was ill-advised and futile (Koch 1935, 1998, Pyne 2001a). In Koch's opinion, developing this wilderness country with roads, trails, fire lookouts, phone lines, and other structures to aid suppression had to a large extent destroyed the special values of a unique and distinctive wilderness area. In the *Journal of Forestry,* Koch's essay was accompanied by a rebuttal from the Forest Service's Washington office. The latter argued that more efficient suppression methods, not a "let burn" approach, were needed and would soon be implemented (Loveridge 1935). Indeed, the Forest Service was ready to raise the ante in its crusade against fire by establishing an aggressive new fire suppression policy (Pyne 1982).

Beginning in 1935, manpower, equipment, and technology were marshaled in a paramilitary campaign against fire under the Forest Service's 10 A.M. Policy. This policy declared that every fire was to be controlled by 10 A.M. the day after it was reported (Pyne 1982). If the control objective was not achieved, then firefighting forces would be mobilized for control by 10 A.M. the following day, and so on. At this time there was only a modest buildup of fuels in many forests because suppression had only been effective for a brief period, and in dry forests grazing had removed much of the grass and other fine fuel. For many years the war on fire was relatively successful. Each year from 1946 through 1978 the total extent of wildfires in the 11 western states was kept below 1 million acres, after having often reached 2 million acres annually between 1917 and 1931 (Fig. 2.1).

The South was the most problematic region for maintaining the fire exclusion policy. Historically, longleaf pine *(Pinus palustris)* made up the most valuable and extensive forests in the South. Detailed research published in 1926 by H. H. Chapman, dean of the Yale School of Forestry, showed that surface burning in these forests was critically important. Frequent burning promoted longleaf pine regeneration, controlled

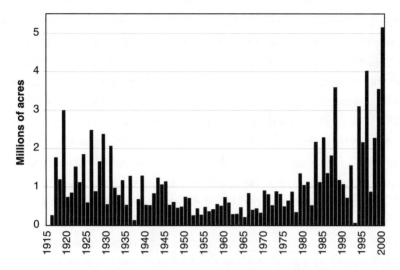

Figure 2.1. Total acreage burned in wildfires annually in the 11 western states. (AZ, CA, CO, ID, MT, NM, NV, OR, UT, WA, WY) 1916 to 2000. Yearly totals prior to 1931 are low estimates because they do not include all land ownerships. Note the upward trend since 1979. (Data provided by the National Interagency Fire Center, Boise, ID.)

brownspot disease in seedlings, and reduced fuels, thus preventing damage from wildfires (Chapman 1926, Pyne 1982). Other southern pines also benefited from light burning. Soon after Chapman published his conclusions, other studies revealed that surface burning in southern forests was crucial for maintaining quail habitat and populations important for the game-bird hunting industry (Pyne 1982). These findings increased pressure on the Forest Service to allow controlled burning in southern forests. The Forest Service had the ability to withhold Clarke–McNary funding from states that engaged in fire practices, such as controlled burning, that the agency did not like. After years of resisting the scientific evidence and pleas from southern foresters and biologists to be permitted to use fire, in the 1940s the Forest Service agreed that controlled burning in southern forests could be allowed. However, the agency made it clear that acceptance of controlled burning in the South was a special case and was not to be extended to the West.

Despite the institutional fervor to vanquish fire, some individual foresters and timbermen in the West continued to speak out in favor of controlled burning. In 1943, Harold Weaver, a forester with the Indian Service in the Department of the Interior, published an article in the *Journal of Forestry* stating his belief that successful management of ponderosa pine forests depends on either finding a substitute for fire as a natural thinning and fertilizing agent or using fire as a silvicultural tool (Weaver 1943). Weaver observed that there had been little success in replacing fire and far too little thought and research applied to using fire as a tool. He argued that removing fire from these forests had serious ecological and practical consequences for maintaining productive forests.

The U.S. Forest Service continued efforts to discredit and ban burning in California timberlands. Top Forest Service officials were not pleased when in the late 1940s, after light burning's apparent demise, a "radical" forester named Harold Biswell at the University of California–Berkeley began experimenting with controlled burning (Pyne 1982). By the late 1960s, Weaver and Biswell, who continued to demonstrate the benefits of burning in ponderosa pine forests, had attracted numerous supporters (Biswell and others 1973).

During the 1950s and 1960s several scientists in the emerging field of ecology also concluded that attempts to eliminate fire in national forests, national parks, and other western wildlands were a grave mistake. A committee of prominent wildlife biologists recommended to the U.S. secretary of the interior that fire be reintroduced in the national parks (Leopold and others 1963). As a result, in 1968 Sequoia and Kings Canyon National Parks began to allow lightning fires at high elevations to burn without suppression (Kilgore and Briggs 1972, Pyne 1982). Soon, similar policies were established in a few other western national parks. By the early 1970s the foresight of Elers Koch had been incorporated into a new program allowing some natural fires to burn in the Selway–Bitterroot Wilderness, and the pioneering work of Weaver and Biswell led to the use of prescribed fire in ponderosa pine forests at several locations across the West.

Foresters Accept Fire, Society Resists: 1978 to Present

In the late 1970s, federal land management agencies acknowledged that they were losing the battle against wildfires (Pyne 1982). Despite increasingly aggressive suppression tactics, costs for fire protection had spiraled upward, growing 8 percent annually since World War II. Even with this increased spending, annual wildfire acreage from the late 1960s through 1977 was higher than at any time since 1945 (see Fig. 2.1). Many large severe wildfires defied suppression efforts, including the Sundance and Trapper Peak fires in Idaho (1967), the Wenatchee fires in Washington (1970), the Carrizo fire in Arizona (1971), and the Marble Cone and Hog fires in California (1977) (Pyne 1982). Also, there was increasing pressure to recognize ecological studies that showed fire had an important role in maintaining some kinds of wildland forests, and especially the large natural areas (Kilgore 1986).

The federal Office of Management and Budget, alarmed by the accelerating expenditures for fire suppression, advised the land management agencies to develop more cost-effective fire policies. As a result, they changed their mission from focusing only on fire control to a broader mission called fire management. This included reducing forest fuels, using prescribed fire, limited suppression of some wildfires, and traditional suppression for other fires (Kilgore 1976, Nelson 1979).

In practice it was difficult to replace the "war on wildfire" with fire management. The Forest Service had campaigned to eliminate fire since 1908, and at first many rural people had resisted, advocating light burning. Seventy years later, forestry was ready to accept the need to use fire in forests, but the public and newly established environmental regulations were more aligned with forestry's old credo that fire is bad and unnecessary.

Just as the 1910 fires had been a defining event that helped launch the national fire suppression policy, the huge 1988 fires in the greater Yellowstone Park area strongly influenced the fledgling fire management policy. Television news dramatized the Yellowstone fires every day for two months. The media often portrayed the fires as overwhelmingly powerful infernos that were destroying Yellowstone's environment (Smith 1992, 1995). The more complex and accurate story of these fires as a natural process received play months later in some newspapers, magazines, and other media, but by then television newscasts and public attention had shifted to other events.

Some commentators and politicians blamed the National Park Service and their "let it burn" policy. Indeed, some of the large fires threatening facilities, communities, and summer homes in the greater Yellowstone area had initially been allowed to burn. When lightning ignited these fires in early summer, they fit within the range of acceptable conditions and were designated as "prescribed natural fires." Yellowstone's natural fire program had been in place since 1972, but the first 16 years had been a period of relatively moist summers, so the management approach allowing natural fires to burn had not been tested in a tough wildfire season (Despain and Romme 1991, Wakimoto 1989). It was well known that historically the Yellowstone high country had burned in stand-replacement fires (Romme 1982). However, these occurred at long intervals of two to four centuries. During more recent summers Yellowstone had had the reputation of being an "asbestos forest," where fires were small or at least nonthreatening. This may have lulled park managers into thinking there was little need to plan for fires of historic proportions.

When the fires of 1988 began to spread, they were fanned far beyond their acceptable boundaries by a series of regionwide windstorms that drove the flames from each fire across tens of thousands of acres in a 24-hour period (Rothermel, Hartford, and Chase 1994). The scale and severity of the 1988 fires in Yellowstone had historical precedent in past centuries, but not in the past 100 years (Millspaugh, Whitlock, and Bartlein 2000, Romme and Despain 1989). However, planners had not prepared for the threats posed by such fires to highly vulnerable facilities, resorts, and homes.

Two hundred miles north of Yellowstone, the 1988 fire season's final windstorm made a direct hit on the Canyon Creek prescribed natural fire in the Bob Marshall Wilderness complex, driving it out of the national forest onto ranches and other private lands near Augusta, Montana (Daniels 1991). This produced still more controversy and criticism regarding the policy of allowing some natural fires to burn. Media portrayals of other record-setting wildfire seasons since then reinforced the message that fire in the forest is inherently damaging and should be eliminated.

Fire managers in the U.S. Forest Service, National Park Service, and other federal land management agencies have made great strides in developing and implementing prescribed burning techniques. Nevertheless, the scale of burning and fuel treatments is only a tiny fraction, probably 1 or 2 percent, of what is needed in most types of wildland forests to maintain historical ecological conditions or to reduce excessive accumulations of fuels. Moreover, constraints on expanding these programs loom large (GAO 1999, Parsons and Landres 1998). Actually making the change in emphasis from fire control to fire management has been elusive. Over the decades, professional and institutional rewards and incentives have been linked to fire suppression (Czech 1996, GAO 1999, Mutch 2001). Firefighters are regarded as local heroes, and their job is to reduce damage that is otherwise certain. The policy revision to integrate preventive management of fuels and use of prescribed fire with fire suppression did not include changes in rewards and incentives. Also, prescribed fire and fuels management is funded at much lower levels than suppression. Those carrying out prescribed burning may be perceived as inflicting damage in the form of smoke, scorching green trees, and risking an escaped burn.

Managing Fuels and Fire or Unwinnable War: The Future

The next few years offer us a chance to adapt to living with fire-dependent forests and to shape fire to suit our needs. The federal government has supported a great expansion of fuels management and use of prescribed fire in western forests, while at the same time markedly increasing suppression efforts. Achieving fire management goals will be challenging and will require considerable public support. In contrast, it will be easy to escalate suppression efforts even though increased spending may be ineffective in averting a continuation of massive wildfires and destruction of homes and facilities. Moreover, the rapid expansion of residential housing and other development into wildland forests will increase the difficulty of managing fire. Fire management can only succeed if forest residents and visitors are willing to adapt to the forest environment where fire is intrinsic, and also willing to allow land managers and fire specialists to manage fire and fuels.

In retrospect, much of our predicament with managing fire in western forests results from imposing a forestry paradigm developed for other

Chapter 3

Fire on the Landscape:
Past, Present, and Future

We can best unravel the relationship between humans and fire in the western United States by reconstructing the historical magnitude and effects of fire on the land. For convenience, we will focus on the average annual area burned prior to 1900, the annual expanse of burning throughout the twentieth century, and trends that suggest the scope of fires in our future. This background can help us evaluate the choice between continuing our present course or expanding the use of fuels management and prescribed burning.

While the West was still a frontier, free-ranging fire was just one of the many powerful forces of nature to which pioneers and other adventurers had to adapt, as American Indians had done for centuries. However, a seminal conflict unfolded once the seemingly limitless frontier had been filled by ranches and small communities linked to growing centers of commerce by railroads and wagon roads. This conflict arose between the untamed landscape of fire-dependent vegetation and a rapidly developing civilization demanding control of its environment. The clash flared during every hot dry summer, a common occurrence in the semiarid West.

Fires in the Past: Widespread Burning

By the summer of 1889, four railroads carried travelers and their belongings across the Great Plains, deserts, and mountains to the Pacific Coast.

A journey that previously had taken six months filled with hardship and peril as pioneer families walked beside ox-drawn wagons now took a mere three to four days (Evans 1990). The immense herds of buffalo were only a memory, and American Indians had been forced onto reservations. The telegraph provided instant communication with the rest of the country, and western cities and towns began to enjoy the telephone and electric lighting. Washington, Montana, and North and South Dakota gained statehood in 1889, and Idaho and Wyoming followed in 1890. The population of the 11 western states had surged beyond 3 million according to the 1890 census, compared to less than 1 million just 20 years earlier.

Suddenly the West was settled and nearly domesticated. Nevertheless, wildfires still burned vast expanses of the country. The hot, dry summer of 1889 provides a fascinating view of the enormous scale of historical fires on the western landscape. That season also exposes the dilemma posed by free-ranging fire to a region rapidly becoming developed and "civilized."

Major John Wesley Powell, director of the U.S. Geological Survey, portrayed the fires of 1889 in testimony to Congress:

> This past season, as an attaché to the Senatorial committee investigating the questions relating to the arid lands, I passed through South Dakota, North Dakota, Montana, Washington, Oregon, and Idaho by train. Among the valleys, with mountains on every side, during all that trip a mountain was never seen. This was because the fires in the mountains created such a smoke that the whole country was enveloped by it and hidden from view (Powell 1891, 208).

Drawing on more than 20 years' experience exploring the West, Major Powell added that "fire in an ordinary year passes over the ground and burns the leaves and cones, etc., only. But there come critical years, five, ten, fifteen, or twenty years apart, critical seasons of great drought, when there is no rain for several months, and the fire starts and sweeps everything away" (Powell 1891, 208).

Powell's characterization of large-scale burning is substantiated by fire history studies covering the region from the Cascade Range in Washington and Oregon eastward across the Rockies to central Montana and western Wyoming (Barrett, Arno, and Menakis 1997). These studies identify 35 episodes, each consisting of one year or a few consecutive years between 1540 and the 1930s when fires burned throughout much of this region. Fire scar dates and fire reports from over 300 locations indicate that the great conflagrations of 1910 were substantially exceeded in regional extent by fires in 1889 and probably also by the fires occurring in about 1869, 1846, 1823, 1802, 1784, 1778, and 1756 (Barrett, Arno, and Menakis 1997). For example, 28 percent more sites across the region recorded fires in 1889 than in 1910 (Fig. 3.1). The 1889 fires are a window on big fire years of the past because at that time in the inland West,

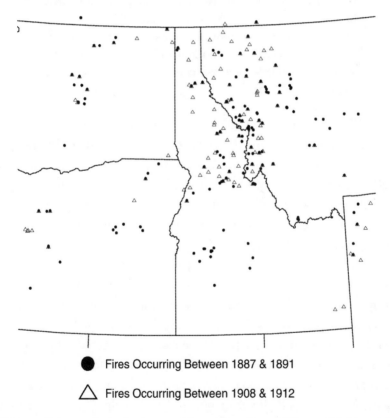

● Fires Occurring Between 1887 & 1891

△ Fires Occurring Between 1908 & 1912

Figure 3.1. Dots mark the locations of forest fires from 1887 to 1891 (circa 1889 fires) and open triangles mark fires from 1908 to 1912 (circa 1910 fires) in the inland northwestern United States as identified in fire history studies. Note the broad geographical distribution and large number of the circa 1889 fires. (Map from Barrett, Arno, and Menakis, 1997.)

the effects of logging were limited and fire suppression was mostly confined to protecting ranches, sawmills, and settlements.

As fires raged across forests and prairies during the summer of 1889, stifling smoke descended on many communities. Although the 1889 fires did not provoke the policy debate that surrounded the 1910 fire season, they did not go unnoticed back East. The *New York Times* reported from Portland, Oregon:

> The atmosphere for miles around is thick with smoke and cinders and burning brands are falling in showers. All the northwestern country seems to be burning up in forest fires. The smoke has been so dense in Portland for the last two or three weeks that for a time it was impossible to see far up the street. . . . In the harbor the smoke has had the effect of fog, and steamers have been required to blow their whistles every few minutes to avoid collisions (*New York Times,* August 15, 1889, 5).

The *Times* also described fires around Miles City, Montana Territory, 1,000 miles to the east along the newly established Northern Pacific railroad: "The bodies of timber north and south of town . . . seem to have caught fire simultaneously. Lightning during the storm of Sunday night is the cause of the fires. The flames approached the ranch of George Rhodes, but the farm hands managed to fight it away. A heavy smoke is hanging over the town" (*New York Times,* August 15, 1889, 5).

Four hundred miles to the west, near the Continental Divide, "forest fires which have been raging all over Montana for three weeks and have destroyed hundreds of thousands of dollars' worth of timber have reached most alarming proportions. . . . In two days the fire traveled over a section 60 miles wide and 100 in length" (*New York Times,* August 20, 1889, 3). Fires burned through millions of acres of mountain forests, but heavily grazed grasslands surrounding several towns including Deer Lodge and Helena, Montana, and Boise, Idaho, apparently helped save them from the flames (Taylor 1989). Extensive areas of the Canadian Rockies also burned, including much of what is now Jasper National Park, Alberta, 270 miles north of Montana (Tande 1979).

Unwary travelers faced considerable danger from free-ranging fires.

On July 30, 1889, two men took the wagon road heading east from northern Idaho:

> They had been warned on account of forest fires raging along the line from there to Thompson [Thompson Falls, MT]. A few hours after they started the roar of the flames was heard, and they urged their teams as rapidly as possible. The speed of the horses was slow compared to the rapidity at which the fire traveled. They were soon overtaken, and leaving their teams in a deep ravine ran for shelter in a deserted tunnel. . . . Their place of refuge was entirely surrounded and it was five days before they were able to get out. . . . There was a small spring in the tunnel from which they obtained water, but they were without food. . . . The flames burned their horses and wagon (*New York Times,* August 14, 1889, 1).

It is arguable that these late nineteenth century fires were more extensive than those before settlement because arriving European-Americans set many fires both inadvertently and on purpose. Indeed, the railroads started and suppressed many fires. However, by the late nineteenth century, heavy livestock grazing in western valleys had removed fine fuels (grass), and American Indian burning was already disrupted in many areas (Arno and Gruell 1983, 1986, Arno, Smith, and Krebs 1997, Baisan and Swetnam 1997). Fire-reducing influences probably offset activities that tended to increase fires. Fire history studies extending back to the sixteenth century or earlier indicate fire frequencies similar to early settlement-era levels (Arno 1976, Barrett and Arno 1982, Barrett, Arno, and Menakis 1997, Fritz 1931, Swetnam 1993).

Throughout this book we use the term "fire return interval," or simply "fire interval," to refer to the period of time between fires at a given place in the forest. This makes obvious sense when the fire interval is shorter than the historical period of time for which we have fire history data. This includes the frequent fires at average intervals of 20 years or so over the last century. But this measure is less intuitive when fire intervals after fire suppression began are longer than the time period we are talking about. An average fire interval of 200 years over the last 80 years of suppression

means that it will be another 120 years before a given place will experience fire again in the future.

A direct comparison of the extent of pre-1900 fires and twentieth-
century fires in the Sierra Nevada of California is quite revealing (McK-
elvey and others 1996). Prior to 1900, the average interval between fires
at any given location varied from 11 years in the ponderosa pine forest to
26 years in the high-elevation red fir forest. It would take 11 years for
areas burned in the ponderosa pine type to equal the area of the entire
type (Agee 1993). During the twentieth century, annual area burned in
the ponderosa pine forest type averaged only about 6 percent of the pre-
1900 average, so fire intervals at a given point would have lengthened to
an average of 192 years (Fig. 3.2). The difference between pre-1900 and

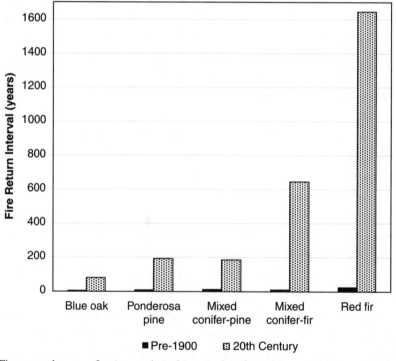

Figure 3.2. Average fire intervals in historical and modern periods in major forest
types of the California Sierra Nevada, extending from the low-elevation blue
oak woodland to the high-elevation red fir forest. (Data from McKelvey and
others 1996.)

twentieth-century fire frequencies was even greater in the red fir forest. This analysis clearly indicates that suppression has had a dramatic effect on fire frequency across the forest landscape. A comparison of historical landscape photographs taken as early as 1849 with modern retakes in the Sierra Nevada shows that fire suppression contributed to dense forest development in many areas (Gruell 2001).

Prior to 1900, an estimated average of 6 million acres burned annually in the northwestern United States (Barrett, Arno, and Menakis 1997). Two-thirds of the burning occurred in dry grasslands and sagebrush. The remaining 2 million acres burned in forest—more than half of that being low-intensity fire in ponderosa pine forests. This northwestern region represents more than one-fourth of the western United States, and if the rest of the West burned at similar fire frequencies, an average of 20 to 25 million acres (forest and grass/sagebrush lands) would have burned annually. This is about ten times as much annual burning as has occurred during the past 20 years. This historical scale of burning is confirmed by Major Powell's observations. From the late 1860s through 1889, the geographical work of his surveyors throughout the West was hampered year after year by heavy smoke during the dry summer months (Powell 1891).

Forests and woodlands in the 11 western states and the Black Hills of western South Dakota cover about 223 million acres, an area larger than all the land in Texas and Oklahoma (Powell and others 1993). The ponderosa pine type historically encompassed about 40 million acres. Other forests and woodlands experiencing relatively frequent fires probably covered another 60 million acres or so, resulting in roughly 100 million acres that was subject to frequent burning. These other frequent-fire types included oak woodlands, some pinyon pine and juniper woodlands, chaparral, some inland Douglas-fir forests, and most forests and woodlands in California (Arno 2000, Küchler 1964, Powell and others 1993, Skinner and Chang 1996). Assuming that average fire intervals were 20 years in the frequently burned areas, which is probably conservative, 5 million acres of these forests would have burned in an average year (100 million acres divided by 20 years). The remaining 123 million acres burned less often, but probably contributed another 1 million acres of annual burning, yielding an estimated average of 6 million acres of forest and woodland burning per year in the West. Western grasslands, sagebrush, and other nonforest types probably burned at least twice the area of forest burning, or over 12 million acres (Barrett, Arno, and Menakis 1997). This produces a total estimate of over 18 million acres of forest and nonforest burning, not far from the 20 to 25 million acre estimate for the West based on expanding the figures from the study of fire history in the northwestern states.

Recent and Future Fires: An Upward Trend

Since the late 1970s, wildfire acreage has increased dramatically (see Fig. 2.1). This is probably related to fuel buildup, regional drought, and a change of policy that has allowed some fires to burn in wilderness and large roadless areas. The year 2000 produced the largest wildfire acreage (over 5 million) in the West since records began in 1916, with about 460,000 acres burned in New Mexico, 950,000 acres in Montana, and 1.2 million acres in Idaho. In the years of large-scale fire activity since 1987, when about 1 million acres burned in northern California and southwestern Oregon, a new fire suppression strategy has gradually evolved. When there are too many wildfires to attack, despite support from national and international firefighting forces, some fires are left to burn because they cannot be controlled effectively and they pose little danger to homes, facilities, and private lands. Concern for the safety of firefighters has recently had a major influence on suppression strategy, especially after the 1994 South Canyon Fire in Colorado where 14 firefighters died. In the 2000 fire season, suppression forces concentrated on protecting homes, ranch facilities, and other developed areas, and fire incident commanders carefully avoided placing firefighters in high-risk situations.

Projecting recent trends into the future, fuels will continue to build up in western forests, costs of fire suppression will continue to escalate, available equipment and trained firefighting forces will be inadequate for suppression needs during severe fire seasons, and concerns for firefighter safety will limit suppression activities (American Forests 1995, GAO 1999, Williams 1995) (see Fig. 2.1).

The population in and near western forests is growing rapidly, as are the numbers and values of homes. Often there is no attempt to reduce forest fuels around homes or to utilize fire-resistant building and land-scaping materials. The large number of homes in the forest serves as a magnet that attracts most of the available firefighting equipment and forces, leaving scant resources available for checking the growth of fires.

The General Accounting Office's analysis of the western wildfire threat details the consequences of fire exclusion policies that have disrupted the normal cycle of fire in many forest types:

> Vegetation has accumulated, creating high levels of fuels for catastrophic wildfires and transforming much of the region into a tinderbox. The number of large wildfires, and of acres burned by them, has increased over the last decade, as have the costs of attempting to put them out. These fires not only compromise the forests' ability to provide timber, outdoor recreation, clean water, and other resources but they also pose increasingly grave risks to human health, safety, property, and infrastructure, especially along the boundaries of forests where population has grown significantly in recent years. [In 1997 the Forest Service] announced its goal to improve forest health by resolving the problems of uncontrollable, catastrophic wildfires on national forests by the end of fiscal year 2015 (GAO 1999, 3).

The General Accounting Office recognized that Congress has indicated support for the Forest Service's initial efforts, but concluded that the Forest Service lacks adequate data and has not developed a viable strategy for tackling this complex problem.

Thus, it appears that the severe wildfires that burned millions of acres in the West in 1988, 1994, 1996, and 2000, while also destroying hundreds of homes, are a harbinger of wildfires in our future. They will produce summers that many westerners will dread. But such a grim scenario can be averted. Considerable scientific knowledge and technology are available to guide reduction of fuels and restoration of more fire-resistant forests, and several demonstration projects are under way. However, before delving into potential courses of action, we need to examine the character of wildland fire itself—the what, why, and how of this powerful force of nature.

Chapter 4

Fire Behavior:
Why and How Fire Burns

The stereotypical news report stating, "Fire has destroyed 20,000 acres of national forest" is highly misleading. Of course, the 20,000 acres are still there—the land has not been vaporized. And the soils have not been sterilized. Less obviously, some forest plants, such as aspen, willows, and many shrubs and herbs, survive even high-intensity burning and resprout from underground parts (M. Miller 2000). Such a news report also gives the impression that all of the 20,000 acres burned uniformly, killing everything. In most cases the acreage reported actually represents the area enclosed by the outer perimeter of the wildfire. This includes significant portions that burned at moderate or low intensity with some trees surviving, as well as some patches that failed to burn at all. During the severe fire season of 1988, about one-third of the area within fire boundaries in the Selway–Bitterroot Wilderness in Idaho and Montana did not burn at all or was moderately burned with numerous trees surviving (Brown and others 1995).

To liberally paraphrase Gertrude Stein, "A fire is not a fire is not a fire." That is, forest fires are so incredibly variable that they almost defy characterization. Nonetheless, fire behavior is the science that attempts to do just that by accounting for the interacting forces that make fire such a dynamic feature of the natural environment.

Fuel

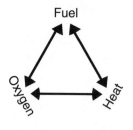

Combustion Triangle

Wildland Fuel

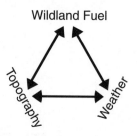

Fire Behavior Triangle

Figure 4.1. The combustion triangle embodies the relationship between the basic elements needed to establish fire. The fire behavior triangle shows the primary factors that influence how a forest fire will burn. (Agee 1993.)

The Combustion Triangle

The basic elements needed to establish fire are fuel, enough heat to sustain combustion, and oxygen. These three ingredients are linked in the "combustion triangle" (Fig. 4.1). Fire suppression involves removing or attenuating one leg of the triangle: digging a fire line to remove fuel, spraying water to remove heat, or covering with dirt to remove oxygen can break the triangle and stop the fire. Using fire in forest management involves selecting the appropriate variables for the fire triangle that will cause the fire to burn in a manner that benefits the land and minimizes risk of damage from the intended fire or a future wildfire.

FUELS

If you've ever tended a campfire or woodstove, you've seen combustion in action. On the chemical level, combustion is the rapid oxidation of fuel that produces heat and gives off gases. When any organic material, be it a piece of leaf litter, a dead branch, or a particle of rotten wood in a decayed log, is heated, any water evaporates and hydrocarbons with low boiling points vaporize. As heating continues, the hydrocarbon gases produced will enter the flaming stage of combustion, oxidizing to release heat energy and combustion products like carbon dioxide. Once the gases have been consumed, flames die out and the remaining carbon material in the fuel begins the process of glowing combustion, or what we see as glowing coals. When this solid carbon has been largely consumed, combustion begins to shut down, normally leaving some charred but unconsumed fuel.

HEAT

Ignition occurs as a result of heating a piece of fuel to its kindling point, or the temperature sufficient to initiate combustion. Development into a self-sustaining fire requires continuity of suitable fuels, such as a carpet of pine-needle litter or a layer of dry grass. As the linkage in the combustion triangle implies, if the moisture content of the fuel is too high, the ignition may be unable to propagate and will flicker out. A breeze that supplies more oxygen to the flames can help the initial ignition to develop into combustion that spreads to adjacent fuels. In forest burning, combustion is usually incomplete, creating carbon monoxide as well as carbon dioxide, and leaving large amounts of charred wood behind (Agee 1993).

Forest fires can be started by volcanoes or even sparks from falling rocks. However, more than 99 percent of fires in the American West are caused by either lightning or human activity. When forestry was being introduced in the United States around 1900, lightning was not generally known to be a major cause of forest fires (Leiberg 1899, Pyne 2001a). Instead, most fires were attributed to human carelessness, which was the case in the eastern United States. This belief probably helped support the impression that fire was an unnecessary and destructive force in the forest. Now we know that lightning is and was a major source of ignition, particularly in the western United States. There are as many as 44,000 thunderstorms worldwide every day. Only a small fraction of lightning strikes cause successful ignitions, but they result in large numbers of fires (Trewartha 1968). An average of about 14,000 lightning fires are reported annually in the United States, and many others die out before being detected.

Relative proportions of lightning- and human-caused ignitions vary from place to place in the western states as well as from year to year, but both are consistently major sources of wildfires. Lightning fires are often thought of as natural, while human fires are regarded as an introduced feature in the forest. However, many studies in anthropology, archaeology, and fire history now indicate that in some of the lower-elevation woodlands and forests, burning by American Indians was a common and wide-spread occurrence long before European contact (Anderson and Moratto 1996, Boyd 1999). For example, in valleys of western Oregon, frequent Indian burning maintained millions of acres of oak woodlands and savan-nas on land that otherwise develops largely into conifer forest (Boyd 1986, Habeck 1961, Johannessen 1971).

OXYGEN

Since oxygen makes up about 20 percent of the lower atmosphere, one might assume that there is always enough for combustion. However, if you've tried to start a campfire or woodstove, you know that it is often necessary to supply oxygen by blowing or fanning just the right amount of air (and oxygen) onto the reluctant, flickering flame. Despite dry fuel, a light draft of oxygen is often needed to nurse the flame for a few minutes, before it will propagate into the carefully arranged fuel bed. More often than not, a lightning spark striking fuel does not produce combustion. Lack of sufficient oxygen can prevent wildland fuel ignition when the spark hits an enclosed area such as deep in the duff at the base of a tree.

The Fire Behavior Triangle

Once a fire is established and burning through the forest, it responds to a "fire behavior triangle" (Fig. 4.1) where the legs are wildland fuels, weather, and topography (Agee 1993). When specific characteristics of each of these factors are known, the behavior of the fire can be predicted, and the way it will burn on the landscape can be modeled to some extent.

WILDLAND FUELS

Many characteristics of fuels, such as type, size, quantity, arrangement, and moisture content, are critical to how a fire burns. Principal forest fuel types include duff (decaying material) and litter on the forest floor, grass and other herbs, different sizes of downed woody material, shrubs, and understory trees. The amount of finely divided dead fuel largely determines flammability, or how easily a fire starts and spreads.

Fine fuels, such as the dead leaves of long-needled pines (ponderosa and Jeffrey pines), dead grass, and small twigs, dry out quickly in the sun. They are easily ignited, and fire spreads rapidly through them. Western forests that historically had an abundance of fine fuels tended to burn frequently in low-intensity fires that fire-resistant trees, protected by thick bark, could survive. Many forests that historically had abundant fine fuels now have sparse fine fuels due to livestock grazing or an increase in the number of short-needle trees such as Douglas-fir. This change in fuels makes it difficult to reintroduce low-intensity fire. Forests with sparse fine fuels are hard to ignite. Dense forests dominated by firs or other conifers with short needles often have little or no grass. Also, the needle litter forms a compacted layer, and there is not enough oxygen to support combustion. As a result, fire occurred historically in dense fir forests only at long intervals under extremely dry conditions and then built up to high intensities, killing many or most of the trees.

Large woody fuels can also be important to fire behavior. For example, 15 to 30 years after a forest has been killed by fire or an insect epidemic, the dead trees will have fallen and partially decayed. Rotten wood ignites and sustains combustion more readily than sound wood because it is often dryer and has many fissures containing air. This abundance of large fuels can contribute to a high-intensity burn that kills all the young trees and also consumes the humus or organic material in the surface soil. Such a high-intensity fire may kill many undergrowth plants that could have survived to resprout after a more moderate fire. Large fuels dry out very slowly, however. If a fire occurs after a short drying period, it may consume most fine fuel and some medium-sized woody fuel, but leave most of the large fuels only charred on the outside. This effect is often a goal in prescribed burning because it reduces wildfire hazard (fine fuels) while maintaining some of the biologically important large rotten wood.

Fuel arrangement is also important in shaping a fire's behavior. Fuels are seldom distributed uniformly throughout a forest. A living forest may have patches of dead, fallen trees here and there, caused by windstorms or insect and disease outbreaks. Between these patches, fuels may be sparse. This uneven fuel arrangement can produce a mixed burning pattern where most trees are killed in some places while most trees survive in other places. The vertical layering of fuels is also critical to fire behavior.

Figure 4.2. The progression of a surface fire to a crown fire via ladder fuels. A surface fire can ascend through ladder fuels, such as thickets of small trees, and burn into the crowns of nearby tall trees. If the ladder fuels were sparse, fire would be unable to crown. (Illustration by Steven Allison-Bunnell/ArtToday.com.)

If there are adequate fuels on the ground, along with an abundance of tall shrubs and thickets of small trees beneath taller trees, fire can climb this fuel ladder and "torch," consuming the foliage of the tall trees (Fig. 4.2). If the trees are dense enough and winds are high enough, flames in the burning tree crowns may spread to adjacent trees in what is called a "running crown fire."

Moisture content of the different kinds of fuels varies and greatly influences fire behavior. When soils are moist, trees and shrubs have ample moisture in their foliage. The weight of their water content is about 150 percent of their dry weight. This level of moisture retards "scorching," which kills but does not consume tree foliage. High moisture content in the foliage also hinders torching, in which fire consumes the crown foliage. However, during extreme drought in late summer, much of the shrub foliage dries and wilts, and even conifer foliage loses a lot of moisture, especially the youngest and oldest of the annual crops of needles on the trees. Crown fire can develop rapidly in these desiccated conditions.

The amount of fuel available at any given time is the net result of several processes that accumulate and deplete fuels (Fig. 4.3). Changes in the

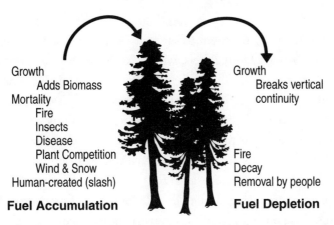

Growth
    Adds Biomass
Mortality
    Fire
    Insects
    Disease
    Plant Competition
    Wind & Snow
    Human-created (slash)

Fuel Accumulation

Growth
    Breaks vertical
    continuity

Fire
Decay
Removal by people

Fuel Depletion

Figure 4.3. The equation governing forest fuel levels. Forest fuels increase and decrease over time as a result of these interacting processes. People can influence these processes by creating fuels such as logging slash or reducing fuels through removal or burning. (Illustration by Steven Allison-Bunnell/ ArtToday.com.)

abundance and arrangement of fuels over a short time can affect how a fire will burn. When live fuels become dead fuels, this increases flammability. For example, perhaps 1 ton per acre of dry pine needles (fine fuels) suddenly appears when the oldest needles die and are cast from the trees in early autumn. Shrub leaves and herbs killed by the first hard frost promptly become a prodigious amount of fine fuel. Before this rapid change, there may not be enough fine fuel continuity to support a fire. A few days later, after the additional fuel appears, fire can spread readily in the same area.

On a longer cycle, fuels change as the structure of the forest changes. Fire behavior will be dramatically different as a forest develops from a young stand of shrubs and saplings to a dense growth of pole-size trees with dead shrubs beneath, to mature trees, and to a stand with dying trees and thickets of younger, small trees. Fire reduces fuels by burning them up, but it can also increase fuels by scorching live foliage and killing trees. This is one reason it may be necessary to conduct two or three successive prescribed burns in some stands to reduce fuels. Similarly, logging can reduce the fuels available for crown fires by removing many tree canopies and making the forest less dense, but it can increase downed fuels if slash is left on the ground.

WEATHER

It is a profound understatement to say that weather has a powerful influence on fire behavior. Unlike fuels, which can be managed, and topography, which remains fixed, weather is the uncontrollable and often unpredictable leg of the fire behavior triangle. For a given amount of wildland fuel in a particular location, weather conditions can make the difference between a creeping surface fire and a roaring crown fire. Prolonged weather patterns such as drought correlate strongly to major fire years in the West (Swetnam and Baisan 1996). Nevertheless, other weather phenomena can muddle the relationship between drought and fire. For instance, during a year with average moisture, massive, dry thunderstorms can produce huge wildfires. Conversely, some drought years have few thunderstorms and, thus, few fires.

Air temperature and fuel temperature influence burning. High air temperatures reduce the amount of heating necessary to bring fuels to the temperature of ignition. Bright sunshine can raise the surface temperature of duff or litter on the ground to 150°F, much higher than the temperature of surrounding air. Sunlight and air temperature act in concert with relative humidity, which falls rapidly with any increase in solar heating, reducing the heat needed for ignition and combustion. Fuel temperature, relative humidity, and thus fire behavior can change rapidly with winds, cloud cover, and migrating air masses. Experienced fire managers watch for the narrow threshold where increasing temperature combined with decreasing relative humidity rapidly cause fire activity to escalate beyond desired levels or beyond control. On the other hand, fire activity may abruptly decrease when a dark cloud suddenly blocks the sun.

Any change in wind speed or direction is likely to have a noticeable effect on a fire. Increasing wind enhances the oxygen supply, speeding up combustion and heat transfer to adjacent unburned fuels. Strong winds may boost fire intensity sufficiently to produce a running crown fire. Wind also carries burning embers that can ignite "spot fires" ahead of the flaming front. During a drought, this is how fires often jump across multiple bulldozed firelines, interstate highways, and rivers. Wind is sometimes essential for successful prescribed burning. To conduct an underburn beneath tall trees, foresters often need a moderate wind to dissipate the heat that under calm conditions would build up and excessively scorch the tree crowns.

TOPOGRAPHY

Fire behavior over rugged topography can change considerably as a fire moves through the mosaic of microclimates on different slopes (Taylor and Skinner 1998, Weatherspoon and Skinner 1995). Steep south-facing slopes are relatively warm and dry, and they support vegetation quite different from that found on neighboring cool and damp north-facing slopes. Similarly, wind-exposed slopes tend to dry out and support more vigorous burning than wind-sheltered (leeward) slopes. In mountainous terrain, daytime solar heating creates upslope and upcanyon winds that tend to drive a fire in front of them. Conversely, nighttime cooling may push the flames downslope or downcanyon. However, the heat from a vigorously burning fire produces an upslope convective wind of its own. Fires burning on steep slopes tend to preheat the fuel upslope from the flaming front. This greatly increases the rate of spread upslope and makes it hazardous to try to suppress the flaming front on steep terrain. Fires advance downslope more slowly. But sometimes burning logs or even burning pinecones suddenly roll down far below the flame front and start spot fires that burn back upslope rapidly to join the main fire. The descending fire front becomes quite irregular and unpredictable.

Types of Fire

Fires are often categorized by the layer of fuels that primarily supports combustion: ground fire, surface fire, and crown fire (Anderson and Brown 1988). Ground fires are mostly smoldering combustion of compacted duff or peat, with little flaming. A ground fire may creep only a few feet per day through the thick forest floor duff under a shady coastal forest. Surface fires burn above the ground in leaves, grasses and herbs, shrubs, and downed woody debris. Crown fires spread through the canopies of trees by torching out trees as the surface fire ignites them from below or by spreading through tree canopies independently of the surface fire. Crown fires are usually driven by strong winds or aided by steep slopes. A wind-driven crown fire can travel many miles until the wind or daytime heating moderates or the fire moves into an area with sparse fuel. In contrast, slope-dependent crown fires only burn to the top of the mountain slope and generally spread down the other side as a surface fire, if at all.

Intensity of a surface fire (the rate at which a fire releases heat energy) can range from very low to very high. In contrast, crown fires are inherently of high intensity. Fire intensity is indicated by flame length, measured along the slant of the flames, which commonly angle upslope or with the wind. As the amount of fuel increases, especially fine fuels, flame length increases and can trigger torching, crowning, and spotting (Anderson and Brown 1988). Fires with average flame lengths less than 4 feet represent low to moderate fireline intensities (less than 100 British thermal units [Btus]/foot of fireline/second) and can generally be controlled using hand tools. Flame lengths averaging more than 8 feet mean much higher intensities (greater than 500 Btus/foot/second) and should not be directly attacked even with bulldozers. High-intensity surface fires are likely to torch and burn tree canopies, which in turn increases intensity and produces airborne embers that can ignite spot fires as much as a mile ahead of the flaming front.

Fire Effects in a Stand

Surface fires are commonly referred to as underburns if a majority of the overstory trees survive. Stands with fire-resistant tree species such as ponderosa pine, Douglas-fir, and some oaks *(Quercus)* historically often experienced low- to moderate-intensity underburns. A similar surface fire in a forest made up of fire-susceptible trees like western hemlock, subalpine fir *(Abies lasiocarpa),* and Engelmann spruce *(Picea engelmannii)* would probably have killed most of the trees, resulting in a "stand-replacement fire." Shallow roots and thin bark make these species vulnerable to any fire. In this case, a low-intensity fire can kill the stand while barely scorching the tree crowns and initially leaving most of the foliage green. If the fire kills some but not all overstory trees, it is called a mixed-severity fire. Often a surface fire will cause stand-replacement burning by heavily scorching the overstory trees even if it doesn't burn up their foliage.

A crown fire completely burns the overstory tree foliage. A surface fire that enters the overhead canopy by climbing ladder fuels and torching becomes a crown fire. Additionally, a canopy fire moving ahead of any surface burning is known as a running crown fire. If you have ever tried to burn a freshly cut conifer branch, you can appreciate that it takes an enormous amount of heat to consume all the moisture-filled foliage in a conifer forest, which can add up to 30 tons of green needles and twigs per acre. Whether stand-replacement fires scorch foliage or consume foliage substantially effects soil nutrition and the potential for erosion. The scorched stand will soon drop a layer of dead conifer needles on the ground, adding nutrients and protecting the soil. In contrast, the stand with no foliage left will leave its exposed soil susceptible to erosion.

Fire Effects on the Landscape

Crown fires give off so much heat, water vapor, soot, and gases that they often produce a massive column of dense smoke (Fuller 1991, Pyne 1984). Rapidly rising superheated smoke can soar to 35,000 feet or higher in the atmosphere, causing a rapid indraft of cooler air to the fire on the ground (Fig. 4.4). This creates great turbulence and fire whirls, or heat-

Figure 4.4. A typical type of convection column generated by large fires. The heat of the fire creates a major updraft. Winds above the surface are stronger than the wind along the ground, bending the column. The strong winds aloft also deposit embers hundreds of yards beyond the fire's leading edge. (Illustration by Steven Allison-Bunnell/ArtToday.com.)

induced whirlwinds. The water vapor from thousands of tons of burning foliage condenses near the top of the convection column, forming a classic anvil-shaped thunderhead, or cumulonimbus cloud, that may even produce a small amount of rain. The gigantic moisture-filled cloud billowing into an otherwise arid, clear sky is awesome testimony to the power of the fire. When a crown fire grows large enough to create its own weather cell, it has reached the status of a "firestorm" or "conflagration" (Fuller 1991).

A strong frontal system with high winds that sweeps across a region containing a large fire will bend the fire's convection column to nearly horizontal. The heat that would otherwise radiate into the atmosphere is blown near the ground and preheats large areas of the forest ahead of the fire front, leading to explosive ignition. The wind-driven fire acts like a gargantuan blowtorch, creating spectacular runs. Hurricane-force winds swept the Big Blowup of the 1910 fires across 3 million acres in a two-day period (Cohen and Miller 1978, Koch 1942, Pyne 2001a). On September 1, 1967, the Sundance Fire in northern Idaho charged 16 miles across a rugged mountain landscape in nine hours (Anderson 1968). In early September 1988, the Canyon Creek Fire in northwestern Montana made a 180,000-acre run driven by high winds. This conflagration sprinted 30 miles in 16 hours eastward over the Continental Divide and onto the Great Plains prairie (Daniels 1991, Devlin 2000). In August 1992, the Fountain Fire in the Cascade Range of northern California became the only fire known to burn all the way from the oak/pine woodlands of the west-slope foothills through the high-elevation forests over the Cascade Crest and down the eastern slope (Skinner 2001). The fire accomplished this in two days, driven by a dry cold front. On the second day, the fire made a run that burned more than 45,000 acres in a few hours, destroying over 300 homes along the way.

To put the power of a huge crown fire into perspective, consider the amount of fuel it consumes: about 35 tons (dry weight) per acre in mixed conifer forest in the northern Rocky Mountains (Brown and Reinhardt 1991). A modern, 2,000 square foot house in a northern climate can be heated for one year with about five tons of dry woody fuel (pellets or firewood). Each acre burned in a crown fire consumes enough fuel to heat seven homes. A 50,000-acre crown fire consumes enough to heat about 350,000 residences.

Less spectacular, but of great ecological importance across the landscape, is the fitful bump and grind behavior that many large fires exhibit over several weeks of burning. Ironically, this common, highly variable fire behavior was not well recognized until the 1970s when some fires in wilderness areas and national parks were allowed to take their natural course for the first time in nearly a century. We have now observed how fires burn over days and weeks of constantly changing weather, moving across rugged, variable terrain covered with a mélange of stand structure and fuels (Brown and others 1995, Taylor and Skinner 1998).

Thinking about Fire Behavior

When news reports state that fire has "destroyed" a certain area, this should pique our curiosity. What is the real situation? What variation exists in the pattern of burning? What proportion of the forest survived? What will the post-fire landscape be like next year? In 10 years? It is always interesting and enlightening to look over the burned area once the smoke has cleared. Some reporters have described the actual effects of fires on the ground, rather than conveying the impression that the whole area was "devastated" (Barker 2000, Devlin 2000). We would like to encourage more of this kind of reporting. The widespread assumption that all forest fire is alike and is uniformly destructive was the rationale that supported our failed fire exclusion policies. If we hope to see fire management policies that restore and maintain desirable forest conditions, we must take note of the contrasting behaviors fire can exhibit and use this knowledge to our advantage in designing management plans.

Chapter 5

Nature's Creative Force:
How Fire Shapes the Forest

Will this burned forest ever recover? How long will it take for trees to come back? What about all the animals? These questions and others floated through public discussions of the memorable fire season of 1988, no doubt prompted by months of national television coverage. Fires in and around Yellowstone National Park were often portrayed as destroying cherished, irreplaceable natural areas (Smith 1992, 1995). After record-setting fire seasons in the 1990s and in 2000, people asked again whether the forests could recover. To biologists and other professionals who study forests, such questions seem puzzling since they know that western forests are well adapted to burning and are actually the product of countless fires occurring over thousands of years. High-elevation forests such as those around Yellowstone Park are adapted to severe fires occurring at long intervals and soon began to recover from the 1988 fires.

The public has been misled. Consider the Smokey Bear posters showing a black, devastated forest with deer and other wildlife standing forlorn, evidently facing starvation, accompanied by the caption, "Fires destroy more than trees." When Yellowstone was burning, elk, bison, and several species of birds often grazed or flocked near the flames, seemingly oblivious to the calamity that riveted the nation. Similarly, advertising campaigns suggest that planting trees is essential to saving burned forests, despite the fact that many western forests promptly grow back to a dense stand of saplings without human assistance. Rather than lacking trees, western forests often have far too many trees as a result of suppression of low-intensity fires (GAO 1999).

The life cycle of a wildland forest is hard to grasp because it transcends the scale of human experience. This community of trees, shrubs, herbs, algae, fungi, soil organisms, insects and other invertebrate animals, birds, fish, reptiles, amphibians, and mammals has thrived on the land for thousands of years. Individuals have lived and died, but these native species have survived countless disturbances from fire, hurricane winds, glacial epochs, and past global warming. We should be concerned about what happens to wildland forests, but perhaps for different reasons than many people think. Rather than being fragile and delicate like fine china, a wildland forest is a living biological community that requires certain disturbances, such as fire, and can tolerate others, such as blowdowns, epidemics, and certain kinds of harvesting or thinning. The biggest threat to wildland forests today is commercial, residential, and industrial development, but it goes largely ignored and undocumented. In Washington State alone, 10 percent of the total forestland, or 2.2 million acres, was converted to various kinds of development between 1970 and 1992 (Washington State DNR 1998). Similar loss of wildland forest is happening all across the West, but it attracts little attention.

While still more "natural" than agricultural lands or plantation forests, wildland forest communities have been significantly altered by suppression of natural fires and affected by logging, roads, forest homesites, and recreational facilities. We cannot simply return natural fires to most of these forests (Parsons 2000, Pyne 2001b). However, if we understand how the trees, undergrowth plants, and animals respond to fire, we may be able to design substitute treatments that can perpetuate the forest community in some approximation of its historical conditions.

Adaptations of Trees

Unlike shrubs and many broadleaved trees, most of the conifers that dominate western forests do not resprout from the stump or roots when the trunk is killed. Redwood is a notable exception because it sprouts vigorously from the stump. Nevertheless, western forest trees do have other adaptations to help them succeed in their fire-prone environment. Trees can be classed as fire susceptible, generalists, or fire dependent.

FIRE-SUSCEPTIBLE TREES

Fire-susceptible trees have little ability to survive fire, and they become increasingly abundant during long periods without fire. They are shade-loving (late-successional) species, including trees of humid regions such as Sitka spruce *(Picea sitchensis)*, western hemlock, mountain hemlock *(Tsuga mertensiana)*, and Pacific silver fir *(Abies amabilis)*. The widely distributed high mountain tree, subalpine fir, is also fire susceptible. These species tend to have thin bark and shallow roots.

GENERALISTS

Generalists are trees that can prosper with or without fires. For example, western juniper *(Juniperus occidentalis)* and other treelike junipers of this region increase in abundance without fire, but they often inhabit environments where fires are relatively frequent (Wright and Bailey 1982). Inland Douglas-fir *(Pseudotsuga menziesii* variety *glauca)* is a widespread tree that colonizes moist habitats as a result of natural fires, but has also expanded into drier habitats as a result of fire exclusion. White fir *(Abies concolor)* and red fir *(Abies magnifica)* similarly prosper with both fire and fire suppression. Engelmann spruce is a high-elevation tree that often seeds in after fires, but it can also remain abundant in habitats that rarely burn. Western red cedar *(Thuja plicata)* often increases with fire suppression, but large, ancient trees of this species have often survived several fires. Redwood prospers without fire, but is also well adapted to surviving fire as well as to regenerating after a burn.

FIRE-DEPENDENT TREES

Fire-dependent tree species have characteristics that allow them to benefit from certain kinds of fire, and they decline in abundance during long periods without fire (M. Miller 2000). Many fire-dependent species are also known as "pioneers" because they are typically among the first to colonize burned or otherwise disturbed sites. Ponderosa pine is the most widespread fire-dependent tree in the western United States. It benefits from frequent, low-intensity fires and virtually assures frequent burning by casting off prodigious quantities of highly flammable needles each year. Historically, frequent burning helped create open-growing stands where tree crowns were generally separated. Most ponderosas have branches beginning high above the ground, and understory trees are scattered. These conditions allowed ponderosa pine to survive surface fires. The frequent burning and open stand conditions often led to development of an uneven-aged, pine-dominated forest. This consisted of clusters and groups of trees ranging from young saplings to immature, pole-size trees; to successively older and larger trees, often including some individuals over 500 years old. Such an uneven-aged pine forest historically occupied millions of acres of land despite the presence of more competitive inland Douglas-fir and white fir (Fig. 5.1). This historical stand

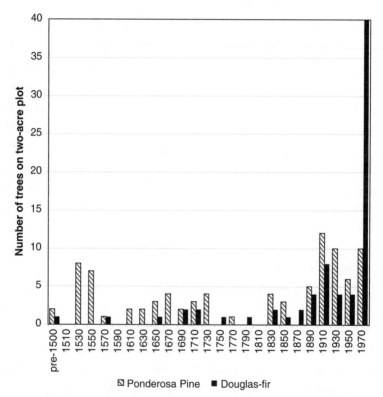

Figure 5.1. The ages (approximate year of establishment) of the trees in an old growth ponderosa pine–fir stand that experienced 11 understory fires between 1630 and 1953. The uneven-aged pattern, with continuing domination by the shade-intolerant pine, depends on frequent fires. (Data from Arno, Scott, and Hartwell 1995.)

structure now seems remarkable because once frequent fire was stopped, an abundance of young firs soon dominated the understory.

Ponderosa pine saplings are superior to firs at surviving historical fires due to their open canopies and large buds protected from fire damage by tufts of long needles. Ponderosa pines produce abundant pitch, which allows them to seal off fire wounds, preventing rot and beetle attacks and allowing the frequently burned trees to survive several centuries—occasionally beyond 800 years (Arno, Scott, and Hartwell 1995). Most of the fire-dependent characteristics of this species are also found in the closely related Jeffrey pine *(Pinus jeffreyi)*. Jeffrey pine occupies a few million acres, mostly in California, and it extends to higher elevations and to colder, more desertlike habitats than ponderosa pine.

Rocky Mountain lodgepole pine is another fire-dependent tree that has several features allowing it to prosper in environments with either low- or high-intensity fires. For example, this species frequently bears some closed cones containing seeds, but sealed tight with pitch, in addition to cones that open at maturity to disperse their seeds (Fig. 5.2). Low-intensity fires usually kill some of these thin-barked trees and allow regeneration to arise from the seeds dispersed by open cones. Seeds sealed

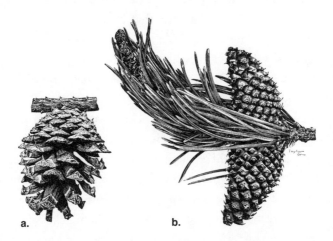

a.   b.

Figure 5.2. Open cone (a) and closed cones (b) of the lodgepole pine. Both types are commonly found in the same natural stands. Closed cones survive severe fires and promptly reseed the burn, while open cones allow regeneration after blowdowns or moderate fires. (Reproduced from Sudworth 1908.)

in closed cones accumulate in the tree crown and remain viable for decades. Closed cones often survive high-intensity fires, and the heat opens the cone, releasing seeds into the newly burned seedbed. Closed cones allow lodgepole pine to regenerate soon after a high-intensity fire, often at the rate of thousands of seedlings per acre.

An even more specialized fire-dependent tree, the knobcone pine *(Pinus attenuata)*, bears only closed cones (Eliot 1948). These are about 4.5 inches long, much larger than lodgepole pine cones, and are very conspicuous in the crowns of the small, misshapen trees. The cones remain closed, but the heat of a fire opens them, dropping seeds into freshly burned soil. Knobcone pine is confined to small areas in the mountains of California and southwestern Oregon.

Mature western larch trees are protected from surface fires by basal bark several inches thick. Coastal Douglas-fir, redwood, and giant sequoia have even thicker bark to protect the sap layer (cambium) from lethal temperatures (greater than 130°F) during a fire (M. Miller 2000, Wright and Bailey 1982). These species typically mature into tall trees with foliage starting high above ground, thus hindering development of crown fire. Coastal Douglas-fir is the ubiquitous fire-dependent tree of the Pacific Coast from northern California to central British Columbia. Historically, it dominated most habitats that burned at short, long, or very long intervals of up to 500 years (Agee 1993). Occasional tall trees survive high-intensity fires. Even when tree crowns burn, fresh cones near the treetop may only be singed and release seed that can be blown widely across the burned area.

In the moist zone of the northern Rockies centered in northern Idaho, western white pine *(Pinus monticola)* is usually the fastest growing tree on burned sites. Rapid height growth enables this pioneer species, often accompanied by western larch and lodgepole pine, to gain dominance in the new forest after fire, despite competition from shade-loving grand fir *(Abies grandis)*, western red cedar, and western hemlock.

Whitebark pine historically occupied large areas of high-elevation forest in the inland northwestern mountains, aided by frequent fires and its seed planter, Clark's nutcracker (Tomback, Arno, and Keane 2001). This noisy, jaylike bird buries the seed in small caches in the soil, preferring semi-open sites like burns. The Clark's nutcracker is the only effective means of dispersal and regeneration for the heavy, wingless seeds, which are almost entirely consumed by squirrels, bears, and other animals when they fall to the ground. High-elevation burns are harsh sites, exposed to strong winds and severe frost; but they are ideal areas for whitebark pine to regenerate. This tree is slow growing, but hardier in an adverse climate than the competing subalpine fir and Engelmann spruce.

Several broadleaved trees of the western forests historically depended on fires to maintain their populations. Most widely distributed of these is quaking aspen, the picturesque, white-stemmed mountain tree whose rounded leaves flutter in the slightest breeze. Aspen trunks are usually short-lived compared to other western trees, but they originate as sprouts from a massive underground root system often occupying several acres. The hundreds or thousands of stems arising from an aspen root system belong to a genetically identical clone, linked together by the root mass. In the West, aspen clones typically appear as small groves, often largely surrounded by sagebrush rangeland or conifer forest. Aspen clones occupy millions of acres in the central and southern Rockies, and they used to be abundant in the northern Rockies. However, throughout much of the West, fire suppression has allowed conifers to outcompete and replace aspen (DeByle, Bevins, and Fischer 1987). Livestock and wildlife relish aspen sprouts, foliage, and bark. This consumption depletes sprouts and kills larger trees when elk and moose eat the succulent bark. Large, high-intensity fires like the 1988 burns in the Yellowstone area sometimes allow prolific regeneration of aspen from its minute windborne seeds. Large burns over 1,000 acres are needed to perpetuate aspen because high populations of elk can consume and kill most of the seedlings or sprouts in small burned areas (Kay 1993).

Several western oaks depend on fire. Oregon white oak *(Quercus garryana)*, which ranges from southwestern British Columbia to central California, and California black oak *(Quercus kelloggii)*, which grows from west-central Oregon to Baja California, readily regenerate in forest openings caused by fire (Agee 1993, McDonald 1990). Competing conifers would eventually shade out both species were it not for periodic fires or other disturbances. These oaks regenerate from stump sprouts and root sprouts when top-killed by fire, and both also produce seedlings from acorns. The moderately thick bark of Oregon white oak provides enough protection to allow widely dispersed oak trees to survive. The oaks formed grasslands with scattered trees, or savannas, in the valleys of western Oregon where American Indians burned annually or every few years. In 1841 the Wilkes exploration party saw extensive fires that "destroyed all the vegetation, except the oak trees, which appear to be uninjured" (Agee 1996, 72). Trunks of California black oak may be more sensitive to fire damage, but as a common associate of ponderosa pine, this species also prospers in an environment where fire is very frequent (Anderson and Moratto 1996, McDonald 1990).

Adaptations of Undergrowth Plants

SURVIVORS

Most shrubs and many herbs in western forests can survive fire because of regenerative buds located on underground stems (M. Miller 2000). Regenerative organs include the root crown at the base of the stem of shrubs and rhizomes, which look superficially like horizontal roots. Fire survivors include shrubby species of maples *(Acer)*, alders *(Alnus)*, willows *(Salix)*, serviceberry *(Amelanchier)*, and manzanita *(Arctostaphylos)* as well as other tall shrubs associated with the chaparral shrublands in California. These shrubs have large root crowns extending deep into the ground (Stickney 1990). Root crowns can resprout if some of their underground buds are sheltered from being heated to a lethal temperature. Most low shrubs, such as *Spiraea*, snowberry *(Symphoricarpos)*, and huckleberry *(Vaccinium)*, and many native herbs, including grasses, have rhizomes. These spreading underground stems may extend down into mineral soil where burning seldom occurs, although heating to a lethal temperature is common. Members of the lily and orchid families have deeply buried bulblike structures that can regenerate following fire. The abundance of underground plant parts that sprout after fire depends on the fire's severity and the depth of lethal heating in the soil (M. Miller 2000). For example, in the strip where a large rotten log lies as it burns, lethal heating extends deep and eliminates most resprouting plants.

Root crowns, rhizomes, and other underground regenerative parts commonly grow at a variety of depths in the soil. This location plus their relative size affects their vulnerability to lethal heating. For example, underground parts of the low ground-cover plants twinflower *(Linnaea borealis)* and kinnikinnick *(Arctostaphylos uva-ursi)* are very shallow, making survival unlikely if a fire consumes most of the soil organic layer (Smith and Fischer 1997). In contrast, rhizomes of the low shrubs *Spiraea* and snowberry commonly survive fires because they are often located more than 2 inches below the soil surface (Bradley 1984). Still, variations in soil heating during most fires allow even vulnerable species to survive in scattered locations. In severe fires, where all the soil organic matter is consumed, most *Spiraea* and snowberry rhizomes are likely to succumb to lethal heating, but some root crowns and other parts of large shrubs like Scouler willow *(Salix scouleriana)*, scrub oaks *(Quercus)*, and manzanitas will usually survive and resprout within a few weeks.

COLONIZERS

Colonizers are another class of plants found on burned sites, although they were not growing there before the fire. Some were hidden in the form of dormant seeds buried in the soil. Plants that regenerate from seed stored in the soil are known as residual colonizers (Stickney 1990). Throughout the West, the most common examples are different species of *Ceanothus*—some of which are called deerbrush, buckbrush, and snowbrush. Their seed coat resists water and heat and protects the seed embryo for 200 to 300 years buried in the soil (Reed 1974). When the dormant seed is heated (but not consumed) by a fire, it germinates. Various species of *Ribes* (currants and gooseberries) are also residual colonizers as a result of seed stored in the soil, as are several herbs, including a species of *Geranium* and wild hollyhock *(Illiamna rivularis)*.

Offsite colonizers also regenerate on burned areas from seed blown in or otherwise transported from unburned areas (Stickney 1990). Offsite colonizers such as fireweed *(Epilobium angustifolium)*, thistles *(Cirsium)*, cottonwoods, and Scouler's willow typically have tiny seeds that are buoyed through the air by attached downy plumes. Many plants, including several introduced weeds, have the potential to colonize a burn this way. Usually a few species of offsite colonizers become abundant and may dominate the new plant community. Then within a few years they are likely to be replaced by undergrowth species that were prevalent before the fire. Offsite colonizers can include the many light-seeded conifers such as western hemlock, western red cedar, and western larch, as well as heavy-seeded trees like several pines and junipers whose seeds or fruits can be brought in by birds and small animals.

The initial plant community the first season after a fire includes survivors from sprouts, residual colonizers from buried seeds, and offsite colonizers from transported seeds (Stickney 1990). The number of survivors depends on how the fire's severity acted on the plants originally growing there (M. Miller 2000). The number of residual colonizers depends on the quantity and survival of seed buried in the soil. Further additions to the new community must come from offsite colonizers, whose success is dependent on a seed crop and a seed-dispersal event that coincide with the availability of the burned site and weather conditions suitable for germination and establishment. Typically, a low-intensity fire will not much change the original undergrowth community, even if many fire-sensitive trees are killed. The initial community will consist primarily of survivors, especially shrubs and herbaceous plants. A severe fire will kill the trees and many underground plant parts, bare the soil, and give rise to a community consisting largely of colonizers.

This pattern can bring problems for maintaining natural diversity when fire exclusion causes a shift to unusually severe fires over large areas. Many historically open ponderosa pine forests previously maintained by frequent low-intensity fires have been replaced by dense thicket stands, and much of the historical undergrowth has been shaded out. In recent years many of these stands have experienced wildfires that left severely burned soil open to colonization by introduced weeds.

Effects on Wildlife

Despite a common perception that forest fires harm wildlife, fires generally kill and injure relatively few animals. The ability of mammals to survive fire depends on their mobility and ability to seek shelter (Lyon and others 2000). Large, highly mobile animals tend to move calmly near edges of actively burning fires. Most small mammals seek refuge in underground tunnels or spaces under rocks. Not all survive, but their populations are high enough that they are seldom threatened with a serious long-term loss.

In contrast, fire's effect in shaping and renewing wildlife habitat is critically important for many species. The large fires of 1988 in the greater Yellowstone area killed only about 1 percent of the elk population (Singer and Schullery 1989). About 40 percent of the northern Yellowstone elk herd did die the winter following the fires. However, this was due to a combination of factors: An already high elk population in relation to the area's normal forage production; drought during the summer of 1988 coupled with loss of forage on burned winter range; hunting pressure adjacent to the park; and harsh winter weather, including a record-setting cold wave that produced wind-chill temperatures of −70 to −100°F (Lyon and others 2000). In the years following the fires, elk grazed burned areas more than unburned areas, and in only five years the population had recovered to pre-fire levels (Knight 1996). Burning of wildland forests generally allows elk populations to increase due to substantial gains in forage production. Enhanced forage commonly follows fires because tree mortality results in more sunlight and moisture for herbs, and the ash provides an immediate fertilizing effect on soils.

Birds are able to avoid injury from fire and smoke, but active nests may be destroyed and fledglings killed if fires burn during the spring nesting season (Lyon and others 2000). Most western wildfires occur in mid to late summer, after nesting season. In southwestern forests, however, fires are most common in May and June. Historically, most of these fires were underburns that would have destroyed nests of ground-nesting birds but probably did not damage many of the nests located high up in trees. If prescribed burns are conducted during the spring nesting season, they can impact nesting birds. Most "spring" prescribed fires are actually conducted soon after snowmelt and before nesting begins. This late winter–early spring period has frequent killing frosts and precedes most new plant growth, which reduces fire's impact on new vegetation as well as on nesting birds.

There are few reports of fire-caused injury to reptiles (snakes and lizards) and amphibians (mostly frogs, toads, and salamanders) even though many of them, especially amphibians, have limited mobility (Lyon and others 2000). Most reptiles and amphibians evidently avoid fire by fleeing or burrowing or because they normally favor wet places in the forest.

Animal species native to areas that have experienced fires over many generations can obviously prosper in habitat shaped by fire. Wildlife biologists were some of the first professionals to recognize the importance of the natural role of fire in forests. Aldo Leopold, the early western forester who helped establish the profession of wildlife management, first argued in favor of fire exclusion (Brown and Carmony 1990), but later identified fire suppression as an important cause of degradation of big game habitat in the mountains of Arizona (Leopold 1924). In the 1930s, research showed that controlled burning improves the quality of big game forage and is critical for maintaining habitat for bobwhite quail in southern pine forests (Lyon and others 2000). Narrow streamside habitats important for wildlife and aquatic communities are now at higher risk of severe burning due to exclusion of fire from the surrounding ponderosa pine and fir forests (Agee 1998).

A blue-ribbon committee selected by Secretary of Interior Stewart Udall in 1963 delivered a ground-breaking report on wildlife management in the national parks (Leopold and others 1963). Headed by Aldo Leopold's son, A. Starker Leopold, the committee recommended an ecologically oriented approach to national park management, including restoration of fire as a natural process. The Leopold report established the concept that wildlife habitat is not stable or preservable in an unchanged condition, but rather is dynamic, and that suitable habitat for many species and communities must be renewed by fire (Lyon and others 2000). The recommendations of the Leopold report were soon considered, not only for management of national parks but also for national wildlife refuges and national forest wilderness. Eventually, the concept of ecologically based management, including restoration of fire, was extended to many national forests and other publicly owned forestlands in the West (Salwasser and Pfister 1994). Today, The Nature Conservancy is leading efforts to restore beneficial fire on privately owned wildlands (Wilkinson 2001).

Just as fire shapes plant communities, animal communities have also evolved in habitats characterized by particular patterns and severities of fire. Animals have developed adaptations not only to fire itself, but also to the changes in vegetation that fire brings about. Animal communities reorganize in response to fire with increases in some species and decreases in others. For example, bark beetles, wood borers, and other insects are immediately attracted to the dead trees in burned forests. Their soaring populations in turn bring woodpeckers and other birds to feed on them (Lyon and others 2000). The black-backed woodpecker, listed as a sensitive species, thrives in burned forests (see Fig. 1.2). Small mammals come to feed on the new plant growth in the first season after a fire, providing abundant prey for predators such as raptors and ample pickings for scavengers too (Lyon and others 2000).

Fires create standing dead trees, or snags, that are critical habitat for many species of birds and small mammals that nest in cavities initially excavated by woodpeckers (Lyon and others 2000). Moreover, many of the most favored and durable snags are from fire-dependent tree species, such as ponderosa pine and western larch (McClelland and others 1979). Interestingly, a fire-killed tree that previously survived many low-intensity fires may stand longer as a snag because of its high pitch content and resistance to decay. A snag that grew up during fire suppression may rot and fall down after only a decade or so, removing that habitat prematurely (Smith 2000a).

Just one challenge to restoring fire is that the management goals for wildland forests can conflict. Planning for protection of threatened and endangered species may not accommodate allowing fire to play its natural role. For example, the northern spotted owl prefers dense old forests. It has expanded its range in ponderosa pine–fir forests on the eastern slope of the Cascade Mountains that have become dense due to fire suppression (Everett and others 1997). Rules made to protect existing owls may try to force preservation of a dense forest stage that is at high risk for severe fire. Other species that also benefit from old, dense forests created by fire exclusion include pine martin, red-backed vole, boreal owl, brown creeper, and Townsend's warbler (Lyon and others 2000). Rather than trying to exclude fire and other disturbance from all of the forest landscape, it may better to restore a more natural mosaic of different stand conditions, including some relatively open-growing forest dominated by large old trees (Buchanan and Irwin 1993, Gaines, Strand, and Piper 1997).

Unlike the northern spotted owl, which prefers dense forest, the Mexican spotted owl, flammulated owl, and northern goshawk need habitat similar to the historical ponderosa pine forests that were shaped by frequent low-intensity fires. Fire exclusion, the resulting dense stands, and the severe wildfires that follow have diminished suitable habitat and contributed to the listing of these species as sensitive or threatened (Illg and Illg 1994, Reynolds, Block, and Boyce 1996).

Generally, young forest communities such as those arising after stand-replacement and mixed-severity fires support a greater diversity of wildlife. No single stage of forest development will provide ideal habitat for all species. Instead, landscapes made up of a patchwork of young, middle-aged, and old forest communities do comprise suitable habitat for the most diverse assemblage of wildlife. Human alteration of natural fire processes has changed patterns of habitats, which will change the way future fires burn, in turn affecting wildlife populations. Perhaps a more accurate slogan for the Smokey Bear poster would be: "Natural fires *create* more than trees. They rejuvenate shrubs, herbs, and habitat for wildlife, including bears!"

Chapter 6

Different Forests,
Different Fires

The year was 1910. The northern Rocky Mountains had experienced unusually warm, dry weather since February. Summer brought oppressive heat and drought. Fields and forests stood parched. By mid August, a black cloud of smoke enveloped everything. Deep within this surreal pall, a wall of flame roared through the thick forests of western white pine, spruce, western red cedar, and other species that blanketed the Clearwater and Coeur d'Alene Mountains of northern Idaho. The heat was so intense and the smoke so stifling that 5 of the 42 firefighters who took refuge in a mine tunnel perished (Cohen and Miller 1978, Koch 1942, Pyne 2001a). After autumn rains in September, the smoke cleared, leaving a thoroughly charred landscape covered with vertical poles that had been living trees.

Fire was also burning forests in the drier Sapphire Mountains 100 miles to the east, southeast of Missoula, Montana. This area had also baked in heat and drought, but the forests were different, and so was the fire. On north-facing slopes, open groves of tall old larch and Douglas-fir alternated in patches with dense stands of small conifers littered with old fallen trees. The fire torched these thicket stands, sending flames above the treetops, but when it entered the groves of tall larch and Douglas-fir, it stayed on the ground and only scorched the lower limbs. Across a narrow V-shaped canyon loomed a steep south-facing mountainside covered with open-growing ponderosa pine and scattered Douglas-fir standing far apart. Here, the fire burned through grass and pine needles as fires had done roughly every two decades in the past. A person wearing modern fire-resistant clothing could have run safely through the low-intensity flaming front.

As reconstructed from historical accounts and fire history studies, two very distinct portraits of burning in 1910 emerge (Cohen and Miller 1978, Koch 1942, Pyne 2001a). Although we do not know all the reasons for the contrasting patterns of burning, it is clear that we can expect to see certain patterns of fire behavior in different types of forests (Brown 2000). For example, stand-replacement burning was the rule in many lodgepole pine forests such as in Yellowstone National Park, whereas low-intensity fires characterized many ponderosa pine forests. Despite all the sources of variation in fire behavior, including fuel characteristics, weather, and forest structure, each forest type within a region historically tended to burn in similar patterns through the centuries. These pre-suppression-era patterns of burning can be divided into three different "fire regimes" based on the frequency of fire and fire's effect on the forest trees. A few kinds of forests had little or no fire, such as the Sitka spruce–western hemlock forests in humid southeastern Alaska; sometimes we say they have a "nonfire" regime. However, in most of western North America the nonfire regime is rare and localized.

At one end of the spectrum, the understory fire regime is characterized by fires that burn through any given place in the forest at intervals averaging between 1 and 30 years (Brown 2000). Forests subject to the understory fire regime consisted of thick-barked, fire-resistant trees growing at medium or wide spacing and having open understories. We would now call the south-slope forest in the Sapphire Mountains that burned in 1910 an understory fire regime.

Historically in the inland northwestern United States, about 40 percent of the forestland experienced the understory fire regime and 20 percent had the stand-replacement fire regime in which most of the trees were killed by each fire (Quigley, Haynes, and Graham 1996). The remaining 40 percent was characterized by a mixed fire regime, where fires alternated between light underburns and stand replacement, or they were of intermediate intensity, killing most of the fire-susceptible trees while fire-resistant trees often survived (Brown 2000). The mixed fire regime had fire intervals of intermediate length and it occurred in areas of intermediate moisture, where forest fuels dry sufficiently to burn during short periods each year. Long-lived fire-resistant trees like redwood, coastal and inland varieties of Douglas-fir, western larch, ponderosa pine, and large western red cedars are the characteristic survivors in a mixed fire regime. Forests shaped by the mixed fire regime are highly diverse and variable, which in turn promotes variability in how fires burn. In the 1910 season, the north-slope forest in the Sapphire Mountains was a mixed fire regime.

The stand-replacement fire regime has infrequent fires at long intervals, averaging between 100 and 400 years (Brown 2000). Moreover, each fire kills most of the trees and paves the way for development of a new forest. Sometimes this burning occurs in large, irregular patches. Moist and cold forests commonly are subject to the stand-replacement fire regime because fuels seldom dry out enough to burn and dense vegetation and dead fuels build up. Eventually, during a drought accompanied by high winds, a fire will develop and burn with high intensity, killing most or all of the trees. This does not necessarily require spectacular flames higher than the tree crowns, as is often shown on television or in movies. A surface fire burning in heavy fuels can have the same lethal effect. The forests of the Clearwater and Coeur d'Alene Mountains that burned in 1910 are characteristic of the stand-replacement fire regime.

As a result of fire suppression and logging, the structure of many western forests has now been altered enough to change the fire regime. Fires are still ignited by lightning and people, but only the most intense fires escape suppression and cover large areas. Most prescribed burns are understory fires, but they cover small areas, only about 10 percent of the extent of the understory fires of past centuries (Barrett, Arno, and Menakis 1997). Today's forests are prone to burn more severely than forests of the past, actually damaging more timber and putting forest homes in greater danger (Fig. 6.1).

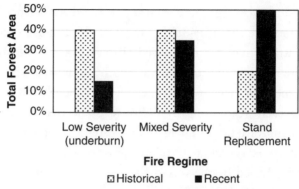

Figure 6.1. The proportion of understory fire, mixed-severity fire, and stand-replacement burning in the pre-1900 period and in recent times in the inland northwestern United States. Note the increasing proportion of stand-replacement fire. (Data from Quigley, Haynes, and Graham 1996.)

Understory Fire Regime

Parklike forests of large, fire-resistant trees characterize the understory fire regime. The stands ranged from being very open to heavily stocked, containing as few as 10 to as many as 150 trees per acre. Usually the understory was open because fires pruned branches high on the tree trunks. Frequent fires also ensured that saplings and tall shrubs were kept relatively sparse. Groves of giant sequoias high up in the Sierra Nevada of California epitomize the understory fire regime (Kilgore 1973). Here, individual trees are so durable and fire-resistant they can live more than 2,000 years through frequent, low-intensity fires; yet without frequent fires the groves become choked with thickets of competing white fir (Swetnam 1993, Weatherspoon 1990). The following contrasting forest types were historically maintained by the understory fire regime.

PONDEROSA PINE

Most of the historical ponderosa pine forest type experienced either an understory or a mixed fire regime. In ponderosa pine forests, 1 to 2 tons per acre of pine needles, branches, cones, and resinous bark flakes accumulate on the ground annually. This mass greatly surpasses the slow rate of decomposition in the semiarid climate that ponderosas occupy. In addition, many of these forests had an abundant growth of grass that dried and was capable of burning every year. Along with these fine fuels, the dry climate allowed burning several months of the year.

Prior to 1900, many ponderosa pine forests were dominated by medium- and large-size pines covered with orange bark divided into distinctive plates shaped like jigsaw puzzle pieces (Fig. 6.2). Due to frequent burning, shrubs, understory trees, and downed logs were sparse (Cooper 1960, Leiberg 1899). Travelers often rode horseback or pulled wagons for miles through these forests without needing to clear trails. In 1857, Lt. Edward Beale wrote of northern Arizona, "It is the most beautiful region I ever remember to have seen in any part of the world. A vast forest of gigantic pines, intersected frequently with open glades, sprinkled all over with mountains, meadows, and wide savannahs, and covered with the richest grasses, was traversed by our party for many days" (Cooper 1960, 129). Pioneers traveling the Oregon Trail through the Blue Mountains of eastern Oregon in the 1850s led their wagon trains through stands of

Figure 6.2. This 1899 photograph of a ponderosa pine forest near Seeley Lake in western Montana predates logging and fire suppression. Frequent understory fires kept most ponderosa pine stands relatively open, with few understory trees and only small quantities of surface fuel. (H. Ayres photo, U.S. Geological Survey.)

magnificent pines and grassy openings, and they noted the smoke from understory fires (Evans 1990, Wickman 1992).

Many of the oldest trees and stumps in ponderosa pine forests have scars that show a consistent pattern of frequent fires dating back as far as the late 1400s (Arno 1976). Just above the ponderosa pine zone in the California Sierra Nevada, there are scattered groves of much older giant sequoias whose fire scars show that the pattern of frequent burning extends back more than 2,000 years (Swetnam 1993).

In the understory fire regime it was difficult for new trees to become established and grow beyond the vulnerable sapling stage. Species that developed the greatest degree of fire resistance at an early age, such as ponderosa pine and western larch, were more likely to survive fires than the fire-susceptible saplings of their competitors, including inland Douglas-fir, white fir, and grand fir (Kalabokidis and Wakimoto 1992). In openings created when overstory trees died, ponderosa pine saplings could grow rapidly and get big enough to resist the next fire. Open-grown pine saplings only 10 years old can survive light burning. Because they regenerated in openings, pines were often distributed in even-aged clusters (Arno, Scott, and Hartwell 1995, White 1985). Ponderosa pine trees might live 400 to 600 years and survive 15 to 30 fires. Nowadays, without frequent fires, ponderosa pines are less likely to live as long or develop to such a large size. Instead, dense stands develop, exposing the pines to competitive stress that results in poor vigor and increased mortality (Arno, Scott, and Hartwell 1995, Biondi 1996, Cochran and Barrett 1998, Covington and Moore 1994a). The fertilizing effect of frequent burning is lost, and a buildup of fuels, especially ladder fuels, makes severe fires more likely (DeLuca 2000).

OREGON WHITE OAK

Groves of oak trees mixed with grassy openings, another fire-dependent vegetation type, grew in the relatively dry valleys and lowlands in the lee of coastal mountains from the "sunshine coast" north of Vancouver, British Columbia, southward to California, where this vegetation covered large areas. Large American Indian populations made these seasonally dry lowlands and foothills home because of the mild climate and abundant resources, including salmon streams and fruit-bearing shrubs and trees, such as oaks.

Archaeological and ecological evidence indicates that natives maintained the extensive oak woodlands of the Willamette and other major valleys by frequent burning (Agee 1993, Boyd 1986, 1999, Habeck 1961). Prior to the influx of European-American settlers in the 1840s, the Kalapuyan people and other tribes typically set fire to large areas of the Oregon oak woodlands to aid hunting and production of food plants (Boyd 1986). Botanist David Douglas describes the aboriginal burning he saw in the Willamette Valley on September 30, 1826:

> Most parts of the country burned; only on little patches in the valleys and on the flats near the low hills that verdure is to be seen. Some of the natives tell me it is done for the purpose of urging the deer to frequent certain parts, to feed, which they leave unburned, and of course they are easily killed. Others say that it is done in order that they might the better find wild honey and grasshoppers, which both serve as articles of winter food (Davies 1980, 94).

In the Willamette Valley, natives commonly ignited the grass in September and burned many areas at short intervals, perhaps annually in some places. Southwestern British Columbia to southern California experienced similar patterns of frequent burning to maintain valley grasslands and open oak woodlands (Lewis 1973, Turner 1999). Most of the fires burned in grass and leaf litter that had accumulated since the previous burn. Oregon white oak in the north, and several additional oak species in southern Oregon and California, survived. Frequent burning heavily thinned sprouts and seedlings of shrubs and trees, while grass and other herbaceous plants flourished (Agee 1993). Frequent fires favored oak, which regenerates from both sprouts and seedlings to outcompete Douglas-fir and brush species. Today, fire exclusion has transformed many places that were once oak savannas into woodlands, while shrub thickets and conifers are replacing oaks in former oak woodlands. Oak woodlands in Wisconsin, Ohio, and Missouri show similar transitions from presettlement times to today; frequent burning by American Indians in the Midwest also once maintained open groves mixed with grasslands (Abrams 1992, Cutter and Guyette 1994, Sutherland 1997).

REDWOOD

Redwood forests cover about 1 million acres in a narrow, fog-bound strip along the California coast from the Oregon border south to Monterey County (Roy 1980). Relatively frequent understory fires were a common feature of the redwood forest, and fires seldom killed many of the large, thick-barked trees (Fritz 1931). How could these damp forests experience fire so often, especially considering that they have a low incidence of lightning? There are short periods, especially in late summer and fall, when there is little rain or fog, and the forests dry out. Once again, American Indian land-use practices shaped the country. Upon Sir Francis Drake's departure on July 23, 1579, from Point Reyes, north of San Francisco Bay, the Miwok people "took a sorrowful farewell of us, but being loath to leave us, they presenly [sic] ran to the tops of the hils [sic] to keep us in their sight as long as they could, making fires before and behind, and on each side of them, burning therein (as is to be supposed) sacrifices to our departure" (Fletcher 1652). The high grassy hills that Drake's party saw being burned are well above the moist redwood-covered valleys, but the Miwoks inhabited both environments.

Redwood is even more fire resistant than its primary associate, coastal Douglas-fir. Furthermore, unlike most conifers, redwood sprouts vigorously from the base when it is top-killed. The most abundant hardwoods in this forest, Pacific madrone *(Arbutus menziesii)* and tanoak *(Lithocarpus densiflorus),* are susceptible to fire. They too resprout when top-killed, but burning stresses them physiologically when they are growing beneath redwood and Douglas-fir trees. Thus frequent fires beneath the towering redwood and Douglas-fir trees tended to hamper growth of hardwoods and create relatively open understories.

Fire scars on the old stumps of redwood trees show that prior to 1850, fire intervals in some of these lowland valley forests averaged between about 5 and 25 years (Brown and Swetnam 1994, Finney and Martin 1989, 1992). Redwood forests in steep mountainous terrain probably burned less often and may represent a mixed fire regime. The record of frequent fires on tree stumps dates back to about A.D. 1300 in one study area (Finney and Martin 1989) and to about A.D. 800 in another (Fritz 1931).

Historical journals, including those kept by Spanish missionaries, corroborate archaeological and anthropological evidence that indicates that aboriginal ignitions made up a large proportion of the frequent fires in some of the redwood groves (Duncan 1992, Greenlee and Langenheim 1990, Lewis 1973, Pyne 1982, Stuart 1987). American Indians could burn during dry periods to maintain meadows to attract deer and elk and to stimulate production of food plants such as wild sunflower *(Wyethia)* and others used for basketmaking such as hazelnut *(Corylus)* (Lewis 1973). Frequent burning kept the forest understory largely free of brush and debris, greatly facilitating travel through what would otherwise have developed into tangles of tall brush. Early European-American settlers in the redwood region adopted the native practice and continued to underburn many of the redwood forests, largely to promote forage for grazing (Pyne 1982). Ironically, in recent decades, with the creation of additional preserves to save the redwood forests, contemporary land managers and environmentalists have shown little interest in returning fire to its primeval role in these ecosystems.

Mixed Fire Regime

The mixed fire regime is notable for the ecological diversity it helped produce, both within a stand and across the landscape (Agee 1998). Overall, the mixed fire regime covers the spectrum of burning patterns from low-intensity surface fires to conflagrations that kill most trees over large areas. This regime also includes many fires that thinned the forest in irregular patterns, killing most of the fire-sensitive trees while more of the resistant trees survived (Fig. 6.3).

The highly variable burning in the mixed fire regime produces highly diverse forests in both structure and species composition, including shrubs and herbaceous plants. Fire intervals averaged between 30 and 100 years, but successive fires sometimes occurred at intervals of less than 30 or more than 100 years apart. Fires in the mixed fire regime are so irregular in frequency and severity that they leave a chaotic, complex pattern

Figure 6.3. This 1899 photograph shows a western larch–lodgepole pine forest north of Seeley Lake, Montana, prior to logging and fire suppression. A mixed-severity fire about two decades earlier evidently killed some of the overstory trees and allowed vigorous regeneration of lodgepole pine (dark saplings) and some larch. (H. Ayres photo, U.S. Geological Survey.)

of evidence on the landscape that was not well described until recently (Brown 1995, 2000). This fire regime became more obvious in the 1970s to observers of lightning fires that were allowed to burn in national parks and wilderness areas. These fires burned at fluctuating intensities as they moved across large areas of forested country during several weeks and even months. Fire specialists witnessed firsthand the remarkable variation in burning patterns of free-ranging fires.

The mixed fire regime produces an incredibly broad assortment of forest types, including mixed conifer forests, coastal and inland Douglas-fir forests, some lodgepole pine forests, and some ponderosa pine and riparian forests.

MIXED CONIFER

Mixed patterns of fire mortality are common in forests that contain a variety of trees having different levels of fire resistance. For example, in parts of the California Sierra Nevada, forests contain extremely resistant giant sequoia, highly resistant Jeffrey pine, moderately resistant sugar pine *(Pinus lambertiana),* and low-resistance white fir (Minore 1979). In the northern Rockies, comparable forests have highly resistant western larch, moderately resistant western white pine and old growth western red cedar, low-resistance lodgepole pine, and highly vulnerable subalpine fir. When a fire of intermediate intensity with flames averaging 4 to 8 feet long burns through a forest like this, the largest, most fire resistant and rot resistant and long-lived trees are likely to survive and prosper; less resistant and durable trees will likely die. The least fire resistant trees are usually more competitive under dense shady conditions, and without occasional fires they develop into crowded stands.

COASTAL DOUGLAS-FIR

Pre-1900 Douglas-fir forests often experienced the mixed fire regime, partly because the coastal and inland varieties of this species together occupy the broadest range of habitats of any western tree. The coastal form of Douglas-fir *(Pseudotsuga menziesii* variety *menziesii)* was the major pioneer species after fires in most of the country along the Cascade Range and westward to the Pacific shore. This distribution includes rain forests that receive 150 inches of annual precipitation and islands in northern Puget Sound averaging only 15 inches of rainfall (Agee 1993). The mixed fire regime shaped drier Douglas-fir forests north of Eugene, Oregon, such as in the Puget Sound lowlands. These forests had large, old Douglas-fir with bark 6 to 12 inches thick, covered with char from the numerous fires they had survived. Even on the poor soil of rocky or hard-pan sites, Douglas-firs often grew to 6 feet in diameter and 500 years of age. In moist areas of Oregon's central Cascade Range, the Douglas-fir forests burned at an average rate of once per century in fires that left a patchy pattern of mortality and many survivors (Fig. 6.4). In drier regions like the Puget Sound lowlands, fires were more frequent (Peterson and Hammer 2001). Apparently, natives set many fires to keep the adjacent small grasslands and oak groves open by killing the invading Douglas-fir seedlings (Norton 1979).

Mixed fire regimes characterized most Douglas-fir forests on both dry and moist sites in California and southwestern Oregon (Arno 2000). East of the foggy coastal strip, a Mediterranean climate with long, dry summers prevails. These Douglas-fir–dominated forests have a very diverse assemblage of species like the mixed conifer forests of the Sierra

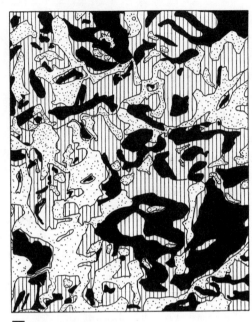

Figure 6.4. Map of a 2.5- by 3-mile area in the Oregon Cascade Range shows the severity of nineteenth-century fires in a coastal Douglas-fir forest. Black patches represent high mortality, striped patches medium mortality, and stippled patches low mortality. Comparable mixed fire patterns were typical of historical fires extending back several centuries. (Reproduced from Morrison and Swanson 1990.)

■ High-mortality patch
||| Medium-mortality patch
∴ Low-mortality patch

Nevada. They were perpetuated by relatively frequent fires of low to moderate severity as evidenced by the abundance of surviving large trees more than 400 years old (Taylor and Skinner 1998, Wills and Stuart 1994).

INLAND DOUGLAS-FIR

Inland Douglas-fir *(Pseudotsuga menziesii* variety *glauca)* also thrives in mixed fire regimes. This seems logical since inland Douglas-fir is the consummate generalist. At one end of its spectrum of diverse habitats, in the humid forests of northern Idaho, inland Douglas-fir is a fire-dependent pioneer. At the other extreme, adjacent to the lower or drought-caused timberline in semiarid high country near the Continental Divide, inland Douglas-fir is the only widespread tree. It was kept from invading mountain grasslands by fires occurring at intervals of about 25 to 50 years. In these intervals, most invading Douglas-fir saplings could not grow large enough (greater than 5 inches in diameter) to survive the next fire (Arno and Gruell 1986). Douglas-firs were able to develop only on rocky or moist sites where fires were less frequent or less intense.

Much has changed in these Douglas-fir/grassland habitats. Beginning in the mid nineteenth century livestock grazing removed grass that had allowed fires to spread. By the late twentieth century, grazing and fire suppression allowed Douglas-fir to colonize large areas of mountain grassland. For instance, the dry northern valleys of Yellowstone National Park support scattered, open stands of ancient Douglas-fir that survived in a mixed fire regime. Some trees attained 4 or even 5 feet in diameter over four to six centuries despite growing on poor sites exposed to a harsh climate. With fire exclusion, Douglas-fir in northern Yellowstone Park and elsewhere has proliferated in young thickets (Houston 1982). In the northern portion of Yellowstone Park, many Douglas-fir thickets were killed in the 1988 fires, restoring a more open, patchy condition.

LODGEPOLE PINE

Rocky Mountain lodgepole pine forests are well known to be associated with stand-replacement fire regimes, as is the usual case in Yellowstone National Park. However, a mixed fire regime occurred in lodgepole pine forests on dry sites in Montana and central Idaho (Arno 2000). Here fuels become dry enough to burn for a few weeks each summer, whereas in Yellowstone fuels become receptive to burning only rarely, under extreme drought. Fires moved through these dry-site forests at vastly differing intensities depending on the time of day or night, the amount of cloud cover, and wind speed. On a calm night, a typical fire backed downhill as a low-intensity underburn that killed few of the thin-barked lodgepole pines. The next afternoon, under sweltering sun and a strong breeze, the same fire would burn at high intensity, killing most trees over hundreds of acres. The next day, under cloudy, cooler, calm weather, the fire burned at a lower intensity, killing trees only where there were concentrations of fuel beneath them.

If you hike for a few miles through these dry lodgepole pine forests, you see the direct evidence of the mixed fire regime. As you trudge along the broad ridges, you pass through a forest mosaic resulting from a single fire a century ago. Along the first part of your route, the fire left scars on the bases of what are now 180-year-old trees, indicating that the fire was a nonlethal underburn. A short distance further up the trail, the fire killed all the trees in a stand-replacement fire, and a uniform young forest of 80- to 100-year-old lodgepole pines has grown up. Farther along your way, patches of old trees, many with fire scars, stand sprinkled through a sea of young trees, indicating a burn of intermediate intensity. In some areas of the forest, patches of still older trees survived two or even three fires (Arno, Reinhardt, and Scott 1993).

PONDEROSA PINE

Historically, some ponderosa pine forests experienced mixed fire regimes, particularly in moist habitats for this species and in areas with steep topography or frequent strong winds. The ponderosa-dominated forests in the Black Hills of South Dakota, on the east slope of the Rockies, and on the upper Great Plains from Colorado to Montana typify these conditions (Arno 2000). Prevalent high winds aided by convection up steep slopes can create stand-replacement burning in places where thick forests develop. This is most common on moist, north-facing slopes. These slopes can also burn in lower intensity fires. Intervening drier or gentler slopes tend to be occupied by more open pine stands and often burn at lower intensity. Thus the landscape had a mixed fire regime that included some stand-replacement burning (Brown and Sieg 1996, Gartner and Thompson 1973, Shinneman and Baker 1997). Many of the ponderosa pine stands in the Black Hills and nearby areas of northeastern Wyoming and southeastern Montana develop dense patches of pine saplings after fire. These become thickets of small, slow-growing trees, susceptible to mortality from bark beetle epidemics and ultimately to stand-replacement fire. Thickets develop particularly on moist sites, whereas dry sites historically had more open stands or only scattered trees.

In the parklike forests of large ponderosa pine in the moist Swan Valley of northwestern Montana, trees of similar age grew up after stand-replacement fires occurring at intervals of about 150 to 500 years (Arno, Scott, and Hartwell 1995). However, in the long intervals between severe fires, these forests had frequent underburns and thus had a mixed fire regime overall. The underburns kept the fire-sensitive saplings of inland

Douglas-fir and grand fir from growing up in the understory and replacing ponderosa pine and western larch. The high frequency of fires may have been due in part to American Indian burning (Arno, Smith, and Krebs 1997).

PONDEROSA PINE RIPARIAN

The narrow stringers of forest bordering streams that flow through the ponderosa pine zone in much of the inland West owe their lush diversity of trees, shrubs, and herbaceous plants to the mixed fire regime. These strips of moist riparian habitat evidently burned at intervals comparable to the adjacent dry upland forests in some areas and somewhat longer intervals in other areas (Arno 1976, McCune 1983, Olson 2000). In places where riparian-site fire intervals were comparatively long (20 to 50 years), fuel accumulation led to mixed fire regimes. Fires thinned the conifers, leaving a moderately open growth of large ponderosa pine, western larch, and Douglas-fir. Broadleaved trees and tall shrubs prospered in the openings. Most of these species sprout after fire and grow rapidly, producing fruit, twigs, or bark that feeds a variety of birds and mammals. In the inland Northwest, fire-dependent broadleaved members of the riparian community included cottonwoods *(Populus)*, aspen, alder, birch *(Betula)*, willows, mountain maple *(Acer glabrum)*, hawthorn *(Crataegus)*, mountain ash *(Sorbus)*, serviceberry, bittercherry *(Prunus emarginata)*, chokecherry *(Prunus virginiana)*, and elderberry *(Sambucus)* (M. K. Arno 1996).

Broadleaved species provide critical nesting habitat for many songbirds, such as warblers and vireos, which migrate north from the tropics each spring (Wheeler, Redman, and Tewksbury 1997). However, with continuing protection from lightning fires and restrictions on prescribed burning, many riparian forests have lost their distinctive character (Tewksbury and others 1999). They are now dominated by conifer thickets, a habitat already abundant across the forest landscape and not heavily used by migrant songbirds.

As areas previously dominated by lush vegetation and less combustible broadleaves, riparian forests have also lost their ability to act as fuel breaks. In recent years many dense conifer stands lining stream courses

Figure 6.5. The aftermath of the 1992 Foothills Fire in the Sheep Creek drainage on the Boise National Forest, Idaho. Modern wildfires commonly burn severely in ponderosa pine forests including riparian areas. (USDA Forest Service photo.)

burned in high-intensity fires (Agee 1998). The old growth pine and western larch have been killed, and water quality and habitat for endangered fish such as bull trout, salmon, and steelhead have been damaged (Fig. 6.5).

Stand-Replacement Fire Regime

The stand-replacement fire regime is common in the high Rocky Mountain forests made up of mixtures of Engelmann spruce, subalpine fir, and lodgepole pine that extend from Colorado to the Yukon Territory. This fire regime also characterizes forests in the coastal mountains from northwestern Oregon to southwestern British Columbia. In the stand-replacement fire regime, fires occur at long intervals averaging 100 to 400 years, and each fire tends to kill most or all trees except at the boundary of the burned area. These fires bring an infusion of diversity as pioneer tree species and the "early seral" species of shrubs and herbs get established and prosper in open, sunny areas, but then virtually disappear as a closed-canopy forest eventually develops.

Stand-replacement fires are commonly large but often gain most of their size in just one or a few short burning periods when high winds drive the flames or when the fire develops a large convection column. The outline of a stand-replacement fire is often quite irregular because of the arrangement of forest fuels on the landscape, features of the terrain, and wind patterns. Stand-replacement burning can result from a lethal surface fire sufficiently intense to kill the trees, or from a running crown fire, which spreads through the forest canopy ahead of flames on the ground.

In rugged mountainous terrain, south- and west-facing slopes and ridgetops are exposed to more solar radiation and often to more drying wind than north- or east-facing slopes. These contrasting exposures usually support different forest types and represent dissimilar burning environments (Arno 2000, Taylor and Skinner 1998, Weatherspoon and Skinner 1995). This heterogeneity may result in a complex landscape mosaic of stand-replacement burning from crown fire and lethal surface fire, interwoven with areas of lighter burning and no burning. Conversely, on gentle topography and more uniform landscapes, such as in the sprawling boreal forests of Canada and Alaska, stand-replacement fires tend to be more consistent, or to at least burn in large patches.

Rather than being purely destructive, the nuances of stand-replacement fire often yield diversity in forest landscapes. The amount of heat the soil experiences during a stand-replacement fire varies widely, ranging from light to severe (Hungerford and others 1991). Nevertheless, the trees die and a new forest of some sort will follow. If soil heating is moderate, most of the shrubs and herbs resprout from surviving underground stems and rootstocks (M. Miller 2000). When soil heating is more severe, a higher proportion of the new plants will probably originate from seed blown or carried in from outside the burn (offsite colonizers), and the new plant community is quite different from the old one (Stickney 1990). Even if the old forest was dominated by shade-tolerant trees like hemlocks or true firs *(Abies)*, the new forest is likely to be dominated by shade-intolerant pioneer trees, such as red alder, coastal Douglas-fir, paper birch, aspen, lodgepole pine, western white pine, western larch, or even whitebark pine. Likewise the undergrowth, which consisted of shade-tolerant plants before the fire, will probably have many disturbance-dependent herbs like fireweeds, thistles, and pearly everlasting *(Anaphalis margaritaceae)* and shrubs like *Ceanothus,* currant, elderberry, and willow (Arno, Simmerman, and Keane 1985, M. Miller 2000).

Western forest types characterized by stand-replacement fire regimes include the moist Pacific Northwest coastal forests, inland moist forests, and cold Rocky Mountain forests (Arno 2000). Whereas forests histori-cally subject to the understory or mixed fire regimes could burn in an average year, forests that historically supported the stand-replacement fire regime would burn only during unusually dry years. A fire may not be able to grow significantly unless there is severe drought and perhaps strong wind. Fires in these kinds of forests are weather-dependent, with fuels playing a secondary role (Agee 1997). For instance, Engelmann spruce–subalpine fir–lodgepole pine forests in the Canadian Rockies accumulate downed fuels and ladder fuels, but only burn in the driest years (Bessie and Johnson 1995). In other kinds of forests, fuel accumu-lation may also be necessary along with drought to allow stand-replacement fire. The fuel accumulation may consist of a layer of understory conifers beneath the main canopy or concentrations of dead and down trees result-ing from a past insect epidemic. It might take one to three centuries for a new forest to become structurally ready to burn again. Stand-replacement fires in lodgepole pine forests generally depend on both weather and fuel buildup (Brown 1975, Knight 1991, Romme 1982).

In some forest types, fuel is the primary factor triggering stand-replacement fire, since weather is generally suitable (Agee 1997). In many ponderosa pine forests, it is dry enough to allow stand-replacement burn-ing for several weeks or longer each year. Historically, however, burning was so frequent that fuels were controlled and an understory fire regime prevailed. Now, a fuel buildup allows stand-replacement fire.

Sometimes two or more stand-replacement fires occur in close succes-sion. The second fire, called a "double burn," happens early in forest development, about 15 to 60 years after the first one. Trees killed by the first fire have fallen into a thick growth of new saplings and tall shrubs. This concentration of fuels near the ground can support a surface fire that delivers more heat to the soil than the earlier stand-replacement burn. This double burn can kill many of the undergrowth plants and remove most of the seed source for conifers since the saplings do not yet bear cones. Double and triple burns in the early twentieth century in Idaho's Clearwater National Forest have created large areas of persistent shrub-fields in what were formerly forested slopes and have caused accelerated erosion in some areas (Barrett 1982, Koch 1998, Wellner 1970). Also, the 1902 Yacolt fire in southwestern Washington and the 1933 Tillamook burn in northwestern Oregon reburned numerous times (Gray and Franklin 1997, Pyne 1982). This long-term loss of forest cover has been a boon to some forms of wildlife. Elk, which were historically scarce in the dense forests of the Clearwater drainage, became abundant amidst the forage-rich shrubfields that followed the double burns. In the Cascade Range, multiple burns on Mount Adams and elsewhere gave rise to productive huckleberry fields sought out by both black bears and people.

COASTAL DOUGLAS-FIR

The cool, wet coastal forests in northwestern Oregon, western Washington, and southwestern British Columbia typify the stand-replacement fire regime. A century ago, Gifford Pinchot described the role of stand-replacement fires in creating forests of magnificent coastal Douglas-fir in what would otherwise have been thick stands of the smaller hemlock, which unlike Douglas-fir is shade tolerant (Pinchot 1899). The greater size and longevity of coastal Douglas-fir allow it to persist in considerable numbers for about 700 years between major disturbances, such as fire or severe blowdowns (Agee 1993).

Scattered individual Douglas-fir survived even stand-replacement fires and served as seed sources in the burns. Seeds may also survive in cones remaining in the crowns of trees that were killed by fire if their foliage was not consumed. The seeds are also wind-dispersed from unburned stands. Douglas-fir seedlings grow rapidly on burned soil, commonly attaining 10 feet of height in five or six years on a good site and then adding 3 to 4 feet in height per year (Hermann and Lavender 1990). This helps Douglas-fir outcompete other conifers in the postburn environment. In some severely burned sites, red alder springs up in profusion and outgrows even

Douglas-fir. Within a few decades, however, Douglas-fir displaces alder, benefiting from soil nitrogen fixed by organisms associated with alder roots. Numerous other shade-intolerant conifers, hardwood species, shrubs, and herbs appear following a fire, greatly enriching the biological diversity of these forests (Fonda and Bliss 1969, Franklin and Dyrness 1973, Hemstrom and Franklin 1982).

A comparably moist "inland maritime" forest centered in northern Idaho also has a history of stand-replacement fires. In this western white pine–western red cedar–western hemlock forest type, white pine, western larch, and lodgepole pine were the principal beneficiaries of stand-replacement fires that occurred at intervals averaging 130 to 300 years (Smith and Fischer 1997). Though this area burns infrequently, the Big Blowup of 1910 shows how much fire can spread in this forest under the right conditions (Cohen and Miller 1978, Pyne 2001a).

LODGEPOLE PINE

In some areas of the Rocky Mountain lodgepole pine forest, a dry climate and quantities of fine surface fuels produce a mixed fire regime. However, much of the lodgepole pine type burns only when there is an accumulation of fallen trees and branches or thickets of understory trees (Brown 1975). When fuels become abundant enough to support fire, stand-replacement burning can occur under extremely dry conditions. This was the case with the Greater Yellowstone fires of 1988, which created a typical mosaic pattern of stand-replacement and unburned areas (Anderson, Ellis, and Romme 1998, Knight 1996). Figure 6.6 shows the irregular patchwork of trees of different ages that often results from a stand-replacement fire in lodgepole pine. Since lodgepole pine is not fire resistant, a lethal surface fire might only scorch the lower branches, turning them ashy brown and leaving the upper crown green. The trees may be so damaged by heat around the base of the trunk that they die within a year or two. A higher intensity surface fire might scorch most of the crown foliage and kill the trees immediately. With still higher intensity, flames would consume the green foliage.

Within 15 to 20 years after a high-elevation lodgepole pine forest is killed by fire or by a bark beetle epidemic (a natural phenomenon in

Figure 6.6. A stand mosaic resulting from historical stand-replacement fires in an unlogged lodgepole pine forest on the Lolo National Forest in western Montana. The young stands in the foreground and left background arose after late nineteenth century fires. The patches of large old trees in the middle distance were young at that time and did not burn. (USDA Forest Service photo.)

mature lodgepole pine) a dense new stand of lodgepole pine saplings develops amid jackstrawed downfall from the former forest (Brown 1975, Romme and Despain 1989). This fuel-rich young stand can easily support a stand-replacement fire. But if the stand escapes burning, by the time it is 100 years old decay may have reduced the abundance of dead fuel, and the trees may have shaded out the lower branches and much of the understory vegetation. This 100-year-old stand is much less prone to burning. A century or two later, when the overstory trees die in a bark beetle epidemic, or when dwarf mistletoe builds up and produces highly flammable witches' brooms in the tree crowns, fire potential goes back up again.

Instead of having a two- to four-century life cycle, lower elevation lodge-pole pine forests may reach maturity and succumb to bark beetles in only 80 to 100 years because the milder climate doesn't limit beetle popula-tions as heavily as colder climates at high elevations.

Fires are critical to maintaining biological diversity in the lodgepole pine forest. Herbs and shrubs are sparse in old stands, but after a fire they soon become abundant. Aspen seedlings reappeared profusely in burned lodgepole pine forests after the 1988 Yellowstone fires (Kay 1993). Black-backed woodpeckers and several other birds, as well as many herbivores, small mammals, invertebrates, and even some aquatic organisms, depend upon fires for producing abundant food insects, recycling nutrients in the soil, and creating productive post-fire plant communities (Despain 1990, Agee 1993).

WHITEBARK PINE

At the highest elevations in much of the mountain West, from central California and Wyoming to southern Alberta and British Columbia, whitebark pine is a fire-dependent forest tree growing with shade-tolerant subalpine fir, Engelmann spruce, or mountain hemlock. Biologists are concerned about the welfare of whitebark pine because in large parts of the Rocky Mountains, grizzly bears rely on whitebark pine seeds as a pri-mary food source (Mattson, Kendall, and Reinhart 2001). Grizzly bears and black bears seek out and rob seeds from extensive cone caches made by red squirrels. In years following a good whitebark pinecone crop, griz-zly bears spend their time foraging in remote high country and seldom conflict with humans by looking for food in campgrounds or around ranches and wildland homesites.

In some areas whitebark pine experiences a mixed fire regime, but this unusual species is most benefited by stand-replacement fires. In most of its historical distribution, which includes 10 to 15 percent of the forested terrain in the northern Rockies, whitebark pine owed its continued exis-tence in past centuries to fires (Arno 1986). Only on the harshest sites, near the tree line, does whitebark pine form pure groves and thus escape competition from subalpine fir and Engelmann spruce. After a large stand-replacement fire, the Clark's nutcracker will transport whitebark pine's pea-sized, nut-like seeds up to several miles and cache them throughout the burn (Tomback 2001). The nutcracker will feed on its seed caches over the following year or two, but many caches are abandoned and will produce whitebark seedlings. Competing tree species rely on wind to transport their seeds, which is less efficient in large burns.

Once established, whitebark pine saplings are so hardy they grow well in high elevations on dry, cold, windswept burns. Eventually the young whitebark pine trees moderate the microclimate in the exposed burn, allowing competing fir, spruce, or hemlock to increase (Arno 2001). Within 200 to 400 years after the fire, the competing species usually crowd out whitebark pine unless another fire intervenes.

Whitebark pine is declining in much of its historical range due to mortality from an introduced fungus, white pine blister rust, and replacement by competing trees as a result of fire suppression (Tomback, Arno, and Keane 2001). Returning fire to whitebark pine communities can counteract this trend and encourage the expansion of genetic strains of the pine that are naturally resistant to blister rust by creating additional open sites where rust-resistant whitebark pine trees could regenerate with the aid of the Clark's nutcracker.

Importance of Fire Regimes

The historical fire regimes that molded different kinds of forests are crucial pieces of information necessary for restoring or maintaining a wildland forest. Over thousands of years, the native plant and animal communities have adapted in numerous ways to the recurrent patterns of fire disturbance. The fire regime indicates which tree species were most abundant and whether overstory trees typically survived fires, fires produced variable mortality, or fires normally killed the trees. Knowledge of the historical fire regime in a given forest helps us resolve whether today's wildfires are similar to fires of the past. We can assess whether a change in forest conditions has brought on a change of fire regime. If modern wildfires fit in with the historical fire regime, we might be able to maintain the wildland forest using natural fires and treatments designed as substitutes for them. If recent fires contrast with the historical fire regime, we might consider restoring a semblance of the historical forest structure and using

Chapter 7

Environmental Impacts: Fire's Influence on Soils, Water, and Air

Whether fire "helps" or "harms" the land is largely a matter of human values and perceptions. When a severe wildfire kills even a large swath of forest, a forest community soon begins to reemerge. Out of the ashes come native plant and animal species that depend on disturbance. In spite of this dependable biological cycle, people often experience fire as an irreversible loss when their homes are destroyed or a favorite recreation area is damaged. Some effects of fire have little impact on the natural system, but can be quite harmful to humans, as when forest fire smoke envelops a city or closes a highway. Fire can also affect natural systems perhaps already stressed by human disturbance such as when erosion caused by natural fire harms a now endangered fish population. Today we simply find some effects of fire, natural or not, to be unacceptable. However, in attempting to avoid undesirable effects of fire, we may inadvertently make matters worse. To avoid smoke we may greatly restrict the use of prescribed fire, in which smoke can be controlled to some extent. This may lead to fuel buildup and a large wildfire, in which smoke cannot be controlled.

Even though we have been conditioned to see large, spectacular stand-replacement fires as purely destructive, in fact they might not inflict any lasting damage to the natural environment. However, many of the stand-

replacement fires throughout the West since 1980 have occurred in ponderosa pine and pine–fir forests that were historically shaped by understory fire regimes. Modern stand-replacement fires *can* degrade these
forests by accelerating erosion, weed invasion, and loss of historical plant
communities or wildlife habitat (Agee 1993, Covington and Moore
1994b).

Conversely, the Yellowstone fires of 1988 were stand-replacement fires
that had minor long-term environmental impacts. Despite news media
coverage suggesting these fires were wreaking havoc on a magnificent
natural area, actual damage was largely related to short-term human aes-
thetic values and risk to recreational developments (Christensen and others 1989, Smith 1992, 1995). Ecologically, the 1988 fires were comparable in scale and severity to fires in the 1700s and earlier (Barrett 1994,
Millspaugh, Whitlock, and Bartlein 2000, Romme 1982, Romme and
Despain 1989). Moreover, the recovery of the land and streams and
regeneration of a new forest at Yellowstone have been impressive and generally prompt (Anderson, Ellis, and Romme 1998, Mahoney and Gresswell 1998, Turner, Romme, and Gardner 1999). The Yellowstone fires
teach us that large stand-replacement fires are normal in some western
forests and that these ecosystems are well adapted to recover from such
fires; indeed, they benefit from fire (Christensen and others 1989). Fires
renewed many native plants, such as aspen, that require disturbance (Kay
1993). Fires are especially important in parks and wilderness, where the
management goal is to maintain naturally functioning ecosystems with a
minimum of human interference.

Fire and Erosion

When a wildfire removes the protective layers of organic matter (litter,
duff, and humus) from the ground surface, heavy rains may seriously
erode the soil (Biswell 1989). "Fire-severity" describes the effect of burning on the soil, and the amount of erosion depends in part on the severity of burning. A severe or "high-severity" fire consumes all of the surface
organic matter (Ryan and Noste 1985). The kind of soil and vegetation,
the steepness of slopes, and the intensity of rainstorms also influence erosion following fire. Historically, soil and streams were damaged in areas

having inherently unstable, erodible soils. This is often the case on steep slopes in the West, especially in areas where the soil is primarily coarse sand from decomposed granitic rock.

While it burns the organic material, a severe fire cooks the underlying mineral soil. After the fire, a water-repellant film often covers the soil surface. The cause of this condition is not fully known, but might result from soil heating or from resins in the duff condensing on the soil surface. Rain cannot soak through this waterproof film. Instead, water washes down the slopes in great quantities, cutting and eroding the soil. In July 2001 typical summer thunderstorms triggered widespread mudflows from severely burned ponderosa pine– and fir-covered slopes in Montana's Bitterroot Valley. Lifelong residents had never seen such erosion, but hydrologists and soil scientists predicted these events based on their analysis of the severity of the fires in 2000.

Buffalo Creek in the Colorado Front Range southwest of Denver is a horror story of erosion and flooding after a wildfire. In what had historically been an open, parklike ponderosa pine forest, fire suppression led to dense stands of small trees and a buildup of litter and woody fuels (Illg and Illg 1997). In May 1996, a wildfire burned 11,900 acres severely, unlike the frequent understory fires the area had experienced in past centuries. That July, a thunderstorm disgorged 2.5 inches of rain in 45 minutes. The fire had created a water-repellant surface on the soil, and it was like releasing the deluge onto an asphalt parking lot. The runoff from burned slopes created a flash flood that demolished the small community of Buffalo Creek, and one man was killed when his vehicle was caught in the flood wave. The quantity of soil lost during a severe fire and rainstorm was measured after the 1984 North Hills Fire near Helena, Montana, was extinguished by a thunderstorm that dumped over an inch of rain (Sirucek 1988). More than 2 inches of soil, about 400 tons per acre, washed off and was deposited at the foot of slopes and in streams. In both cases, the fire's aftermath was not only dangerous to people, but genuinely damaged the landscape.

However, even when large fires burn through heavily fueled forests in erosion-prone terrain during extreme drought, they might not cause damaging erosion. In August 1987, during an extreme drought, wildfires burned nearly a million acres of the steep mountain country of northern California and southwestern Oregon (Reider 1988). An enormous mass of dry lightning storms pounded this region with 12,000 lightning strikes and produced raging wildfires. If severe burning had continued, heavy winter rains later would have eroded the steep slopes, unleashed mudslides, and filled streams with sediments. This could have damaged the habitat of threatened and endangered varieties of salmon and steelhead trout.

Soon, however, a weather and smoke phenomenon common to these mountains moderated the fire behavior. A huge high-pressure area stalled over the area for several weeks, causing the smoke to build into a smothering cloud. The resulting major cooling subdued the fires. Most subsequent burning was not as severe, and there was little damage to the vulnerable soils (Weatherspoon and Skinner 1995).

Mountain slopes sometimes have layers of clay or other impermeable material a few feet beneath the surface where water from heavy rains concentrates, producing an unstable slippery layer. Tree roots stabilize many inherently weak slopes, so when the trees are killed, mud and earth slides are likely. Some of the mountains surrounding the Los Angeles basin are notoriously unstable for this reason. Severe fires in the brushland (chaparral), oak woodland, and ponderosa pine forest followed by heavy rains often trigger catastrophic erosion (McPhee 1988). These mass movements can send a million tons of water-saturated debris, including giant boulders and mudflows, rumbling through mountainside residential areas, causing mass evacuations, casualties, and hundreds of millions of dollars of property damage.

Some unstable areas like the Clearwater River canyon of northern Idaho have experienced severe fires and accelerated erosion for centuries (Barrett 1982). Other unstable areas historically had frequent understory fires, but recent stand-replacement fires have caused much greater erosion. Along the Ponderosa Pine Scenic Highway (Idaho 21) in central Idaho, you can readily observe erosional deposits of rocks and gravel where many of the rivulets and small streams empty into the South Fork of the Payette River near Lowman, northeast of Boise (Fig. 7.1). This erosion occurred after a severe fire burned the ponderosa pine forest in 1989 (Fuller 1991).

Figure 7.1. Accelerated erosion along a small stream descending through a severely burned ponderosa pine–fir forest near Lowman, in central Idaho. Historically these forests burned at intervals of 10 to 30 years in understory fires that probably resulted in much less erosion than the severe fires of the late twentieth century. (Stephen Arno photo.)

Fire and Stream Habitat

Severe fires followed by heavy runoff flows can harm streams and their aquatic food chains (Biswell 1989, Minshall and Brock 1991). Many fish require relatively clear water, and their habitat may suffer when gravel spawning beds silt up and the water becomes muddy. Severe fires also raise stream temperatures by removing shade, which can reduce survival of native cold-water fish like cutthroat trout. Under natural conditions, the effects of severe fires on fish populations were probably mitigated by the ability of the fish to migrate back into damaged areas once the habitat had begun to recover. Today, some populations of native bull trout, west-slope cutthroat, salmon, and steelhead have been reduced and fragmented by roads, dams, mining, logging, and many forms of land development. Some of the more vulnerable fish populations are now restricted to small, isolated segments of streams. If a wildfire and subsequent erosion damages their remaining habitat, some of these populations may have nowhere to go and will not be able to recover. This problem is compounded by fuel buildup in forests formerly in understory or mixed fire regimes. If these forests burn in severe wildfires, damage to stream habitats is likely (see Fig. 6.5 and Fig. 7.1) (Agee 1998). On the other hand, when modern fires burn in a manner comparable to historical fires, permanent damage to the aquatic system is less likely, as seems to have been the case in the aftermath of the 1988 Yellowstone fires (Minshall, Robinson, and Royer 1998).

Fire and Soil Fertility

Nitrogen and other important nutrients held in the duff and litter layers on the ground surface are largely unavailable to plants (DeLuca 2000, Hungerford and others 1991). These organic layers are often too dry in summer and too cold in winter to facilitate decay or plant root development. This in turn prevents plants from accessing the stored nutrients. A severe fire consumes most of the nutrients, although it releases some and makes them available to plants. However, a fire that partially consumes the organic layers releases a flush of nutrients into the soil while the remaining charred duff continues to break down, providing additional soil enrichment over the next few years. Historical fires converted litter and other fuels to ash, which counters the excessively acidic conditions of many soils beneath conifer forests. By releasing nitrogen, potassium, and other nutrients, fires enhance the growth and nutritional value of plants eaten by wildlife, and stimulate flowering and fruit development. If the fire kills and thins the trees, a more diverse and productive community of herbs and shrubs can develop. This in turn enriches the soil. For the first few years after a prescribed burn in a ponderosa pine forest, lupine, arnica, pinegrass, and a variety of other native plants often flower profusely in response to the nutrient flush and increased sunlight. On an adjacent similar tract that was not burned, the same species are present but seldom flowering. Also, after fire, new seedlings of ponderosa pine, western larch, and other conifers often exhibit more luxuriant growth and a deeper blue-green color than seedlings on adjacent unburned sites. In moist forests, fire was infrequent but still influenced soil development. Even fires at intervals of 200 to 400 years helped determine what kind of vegetation occupied the land, and the vegetation affected soil development.

Still, even low-intensity fires burn some of the litter and duff, thus consuming and vaporizing some of the nutrients held in these organic layers. This raises the question of how soil nutrition was maintained over long periods of time in forests such as ponderosa pine that experienced frequent low-intensity fires. These processes are poorly understood, and there are few if any forests to study that have not been affected by exclusion of fire (DeLuca 2000). However, it seems likely that frequent low-intensity fires impact soils in these forests less than do long periods of fire exclusion and a buildup of organic matter followed by a severe wildfire. Fires that consume all the forest floor organic matter also heat the soil, changing it physically and chemically. A high-intensity crown fire or even an unspectacular surface fire that burns thoroughly over many hours or smolders in thick duff for a few days can significantly affect the soil by overheating it and removing most of the nutrients stored in the duff.

Forests that historically burned in stand-replacement fires have developed plant communities and soil organisms that are able to restore a new forest with surprising speed and efficiency (M. Miller 2000, Stickney 1990). Because of their historically long fire intervals, fire suppression has not yet altered many of these forests. Nitrogen-producing pioneer plants, such as lupine and other legumes, alder, and *Ceanothus,* all contribute to the rapid recovery process (M. Miller 2000). The recovering plants are assisted in taking up nutrients by a kind of soil fungi that associate with plant roots (mycorrhizae), which thrive on charred wood in the soil left behind by stand-replacement fires (Graham and others 1994).

To avoid damaging the soil, prescribed fires are often planned for a time when the lower part of the duff and the underlying soil are still moist. The goal is to leave some charred duff, which is good for soil nutrition and structure, and to avoid heating the soil excessively. It is possible that ecological effects of prescribed fire may be different from historical fires because the prescribed burning is often done at a different time of year, such as early spring. Such concerns need to be weighed against the probable effects of continuing to exclude fire and effects of high-severity wildfire linked to fuel buildup. There is simply no way to immediately return to the historical conditions and processes in many western forests once their natural fire regimes have been altered. Using prescribed fire and other fuel reduction measures can help restore more natural fuel situations, which could then allow restoration of natural fire or a closer substitute.

Maintaining Native Plant Communities

The native plant communities in western forests were shaped by the historical fire regimes. The disruption of fire regimes in many forests has had a major effect on the native vegetation. Prolonged fire exclusion has reduced habitat for fire-dependent herbs, shrubs, and tree species. Severe wildfires related to fire exclusion can harm native species that depend on frequent low-intensity fires, such as ponderosa pine and Idaho fescue (Agee 1993). Introduced plants that vigorously compete with and displace natives complicate the situation. Weeds are especially problematic in the warm, dry forests. Scotch broom (Cytisus scoparius) west of the Cascades and knapweeds (Centaurea) inland now occupy millions of acres.

In moist or high-elevation forests that historically had stand-replacement fires at long intervals, fire exclusion can reduce the proportion of the landscape occupied by disturbance-dependent native plants. These communities have a rich assortment of pioneer species and important features for wildlife habitat, such as berry-producing shrubs. To some extent young communities that develop after clearcutting or other heavy logging resemble natural fire–initiated stands (Kauffman 1990, Means, Cissel, and Swanson 1996). However, heavy equipment and logging activities plow up and compact the soil, which allows introduced species and natives such as red alder to colonize logged areas. These sites develop different plant communities than those created by fire. Harvested communities also lack the remaining burned trees, and they have different soil conditions. Moreover, they are often planted in a virtual monoculture to rapidly produce a new tree crop. In contrast, natural regeneration in a burned forest often brings out a diverse assemblage of native herbs, shrubs, and trees.

You might expect that native plant communities would thrive in the large natural areas across the historical stand-replacement fire regimes. However, many of these areas primarily found in national forests and national parks are protected from fire. The overall result is that natural fire–initiated communities are increasingly scarce (Agee 1993). These communities include many pioneer species of grasses and other herbs, important fruit-bearing shrubs, and trees like aspen, oaks, and whitebark pine.

In drier forests that historically had understory fire regimes, the effects of fire exclusion and fires when they do occur are varied and complex. Fire exclusion has created many forests filled for decades by an unusually dense growth of conifers that shade out much of the native undergrowth. The surviving remnants of undergrowth plants that depend on frequent fires have now missed several natural fire cycles. In some areas livestock grazing has further depleted native plants. Deer and elk have heavily grazed the remaining native flora. Introduced plants have become established on logging roads or other disturbed areas and are spreading rapidly.

Under these circumstances, restoring a semblance of the natural plant community involves applying prescribed fire and other restoration treatments in a manner that will reduce impacts on stressed native plants while still gaining the benefits of treatment. Continuing to exclude natural fire and not carrying out substitute treatments might maintain the status quo for native plants and weeds for an unknown period of time. However, inevitably severe wildfires in dense stands with impoverished natural undergrowth and introduced plants are likely to allow weeds to spread and perhaps gain dominance (Agee 1996, Arno 1999, Tveten and Fonda 1999).

In these forests, prescribed fire must not harm the weakened native plants. These plants are often vulnerable to fire damage because they are now growing in unusually deep duff. It is important to plan the season of burning in relation to the annual growth cycle of plants. If fire is applied when plants are actively growing in spring, they are likely to sustain more injury than would result from burning when plants have completed new growth or are largely dormant (M. Miller 2000). In dry inland regions, this dictates burning in late winter when the snowpack melts or in late summer or autumn.

Trees should be thinned frequently to allow native plants to prosper, but ground disturbance by equipment needs to be minimized to reduce weed invasion. Carefully targeted herbicide spraying of certain introduced species can help restore a community of native plants (Manning 2001, Rice 2000).

Rehabilitation to Prevent Erosion

Rehabilitation efforts after major wildfires often rival the urgency, scale, and expense of the fire suppression effort. Wildfire rehabilitation is aimed at reducing soil erosion and damage to streams, gullies, and roads. In burned areas where soils are not highly erosive and where the burn impact on the soil was moderate, most of the rehabilitation effort focuses on reclaiming firelines and protecting roads from increased runoff by installing larger culverts. Bulldozed firelines are erosion-prone because they are built hurriedly and are not planned and constructed to handle runoff safely. By the time they are controlled, large wildfire complexes may have a few hundred miles of bulldozed firelines, many of which are steeper than any roads. These fireline networks probably would not pass an environmental analysis or environmental impact statement, but as emergency measures they are not subject to environmental review. After the fire, firelines need to have cross ditches or waterbars installed for drainage before heavy rains begin. Often the tops and sections of the trees cut in making the fireline are laid back across the fireline to reduce erosion.

One of the ironies of excluding fire from most wildland forests (to protect the forest) is that when a wildfire does occur, we may inflict worse damage to the soil and watershed by bulldozing firelines. If we used prescribed fires instead to manage the forest, we could plan and dig narrower, less damaging firelines by hand. On steep slopes, even a hand-dug line the width of a narrow hiking trail needs to be cross-ditched or water-barred to divert runoff.

Annual grasses or other cover crops are commonly seeded on bulldozed firelines with species that are likely to die out as the natural vegetation redevelops. Native grass seed is an increasingly available alternative. Aerial seeding of grasses across large areas of wildfires on steep terrain is often unnecessary and may even be counterproductive (Crane, Habeck, and Fischer 1983, Schoennagel and Waller 1999). If native plants can resprout from rhizomes or return from wind-dispersed seed, as with fireweed, there is no advantage in planting grass. Seeded introduced species have a competitive advantage over native species like most conifers that slowly seed in from outside the burn. Introduced grass may grow profusely, cure in late summer, and itself become a fire hazard within a couple of years.

Felling selected fire-killed trees, laying them level across the slope, and staking them in place can help reduce slope erosion. Soil must be raked against the downed trunk to fill in any gaps. This is expensive and may not be necessary or at least not as important as protecting burned-out gullies from runoff flows (Miles, Haskins, and Ranken 1989). In erosion-sensitive areas that experience severe burns, check dams made of straw bales or more permanent log or rock structures are often constructed in draws and gullies to slow the runoff (Agee 1993). Straw is sometimes spread in windrows across the slopes, and it is available in compact strips called wattles. Windrows and wattles have been used especially to protect homesites situated directly below severely burned forests. All of these measures are designed to reduce damage for a few years until the natural vegetation recovers.

In the longer term (15 to 60 years) severe erosion could result if a second fire or double burn occurs when the dead trees have fallen and added a heavy concentration of woody fuels to the dense new stand of shrubs and small, tightly limbed trees (Brown, Reinhardt, and Kramer 2001, Gray and Franklin 1997, Koch 1998). To reduce this risk it may be possible to remove a significant fraction of the dead timber in a commercial salvage operation. This should be done with minimal impact to the soil, such as on winter snowpack or using a helicopter. Removing entire dead trees, not just extracting commercial logs, is necessary for fuel reduction. The tops and limbs can be burned in piles at a landing. Some large trees should be left as snags to provide habitat for cavity nesting birds and small mammals (McClelland and others 1979).

Fire and Air Pollution

Managing air quality in large regions of fire-dependent forests is extremely challenging. Under the 1990 Clean Air Act, federal and state air regulators must maintain good air quality for public health and welfare. Carbon monoxide and fine particulates less than 2.5 micrometers in size are the pollutants in forest fire smoke most harmful to human health

(Agee 1993). People who have breathing disabilities such as bronchial diseases, emphysema, and asthma are especially vulnerable to the particulates in smoke. The Clean Air Act protects air quality, but ignores any influence of the regulations it mandates on the health of forest ecosystems (Coloff 1995). Furthermore, regulations limiting regional haze are purely motivated by aesthetics and have no direct impact on human health. They do not account for the fact that, historically, ongoing fires typically made western skies hazy (Evans 1990, Gruell 1985b, Pyne 1982).

Large wildfires often produce gigantic quantities of smoke that can settle over populated areas for days or weeks, far exceeding maximum daily pollution levels produced by large cities or industrial complexes (Devlin 2001a). The 1987 fires in northern California and southwestern Oregon created extreme smoke pollution. Soon after the fires blew up and were burning on hundreds of thousands of acres, stagnant air under a high pressure system settled over the region (Robock 1988). A smoke cloud blocked the sun, causing a temperature inversion that trapped cold air beneath the thick pall. Cooler temperatures, lack of both sunshine and wind, and higher humidity moderated the fire behavior. Lower intensity burning continued and created more smoke, constantly blocking the sun and intensifying the temperature inversion. As a result, daily high temperatures at Happy Camp in the Klamath River canyon were more than 27°F below normal for a week and more than 9°F below normal for three weeks.

During the worst of the inversion, 400 people a day in the sparsely populated Klamath River canyon were treated for respiratory problems. Being outdoors was rated equivalent to smoking three packs of cigarettes a day (Hull 1987). Some days the smoke was so bad that flashlights were needed at midday to read maps, and firefighters would sprawl out in an oxygen tent called the "fresh-air lounge." During the 2000 fire season, a similar weather phenomenon and regional smoke cloud produced record-setting air pollution in western Montana and central Idaho, often reducing visibility to about 100 yards.

Until recently, air quality regulation seemed to exist in a policy vacuum (Mutch 2001); controlling one environmental problem like forest fire smoke was not recognized as contributing to other environmental problems, such as the danger of catastrophic wildfires or loss of biological diversity (Haddow 1995). Regulators and legislators largely ignored the fact that the western environment inherently produces episodes of smoke pollution from forest fires. Unwanted wildfires were considered acts of nature and thus beyond the purview of air quality regulations. Air regulators had no responsibility for wildfire smoke, even though it has produced some of the worst regional air pollution episodes in the West, and even though fuels management and prescribed fire could make such episodes less dangerous and unpleasant (Devlin 2001a, Ottmar, Schaaf, and Alvarado 1996).

During the 1990s, natural resource managers and air regulators began to work together on national, regional, and local levels to develop a more comprehensive approach to maintaining acceptable air quality while acknowledging the reality that western forests will burn. Air regulators now recognize that carefully managed prescribed burning and other fuel reduction activities are a logical part of air quality management (Juvan and Habeck 2001). Land management agencies in the Chelan/Douglas County area of central Washington have cooperated with state air quality regulators to develop an action plan including prescribed fire as a prevention or mitigation measure for wildfire (Washington Department of Ecology 1997). In addition, state authorities in Oregon and Washington have adopted special provisions for prescribed burning for forest health restoration purposes (Core and Peterson 2001).

One of the challenges in harmonizing fire with air quality regulations is getting all the individuals who burn to abide by the regulations. In many rural areas there are virtually no consequences when private individuals violate burning rules. During autumn temperature inversions when smoke is trapped in populated valleys, open burning is officially shut down, but some people continue to burn and produce significant amounts of smoke pollution.

Managing Fire's Impacts

Most foresters, fire managers, and field biologists have abandoned the belief that fire can be eliminated from western forests. Yet many of our fire suppression activities and rules aimed at protecting the environment are still rooted in that paradigm. This unrealistic approach currently prevents managers and citizens from developing good management alternatives that will ultimately be more successful in protecting the soil, plant communities, wildlife habitat, and air quality most people expect in western forests.

Mitigating the effects of severe fires after the fact will always be harder than preventing them from being so severe in the first place. All the post-fire restoration efforts available cannot stop the erosion in large areas, such as the 2001 mudslides in Montana's Bitterroot Valley. Contentious salvage logging over thousands of acres would not be proposed if the fires did not burn such large areas at high intensities. And dangerous smoke pollution from uncontrollable wildfires would not choke our communities if we would tolerate smaller amounts of smoke from prescribed burns on a more predictable basis. Even though fire is itself an inexorable force of nature, we need not view its worst effects as inevitable.

Chapter 8

Fire History: Discovering Effects of Past Fires in a Forest

Perhaps you are a woodlot owner and you would like to know which tree species were abundant in the original forest before any logging. How large, old, and far apart did the trees grow? Did they survive fires, or were the trees typically killed? Did a new stand regenerate en masse after fire or some other disturbance? Or did the forest regenerate continually with saplings replacing individuals or groups of dying trees? It is often possible to answer these and similar questions using such common forestry tools as a chain saw and an increment borer (the hollow drill used to take a core sample of a tree's annual growth rings). Reconstructing the history of a forest provides valuable insights for forest management by revealing how historical fires and other natural forces shaped the forest in the past. You can then compare historical conditions and their effects on forest structure with more recent conditions including fire suppression and logging and their effects on forest structure. To see how this can be done, here is a portrait of a forest that no longer exists except in the traces left behind by ancient stumps, fire scars, and age classes of trees.

A Bygone Forest

The year is 1900. High on a steep mountain slope overlooking a large grassy valley stands a forest of mature ponderosa pine and western larch trees growing widely spaced, nearly 25 feet apart on average. These trees

103

have high, open canopies with no branches on the lower 30 feet of their trunks. The pines are clad in thick orange-brown plated bark, and the larch bark is similar, but with a tinge of purple. Splotches of black char cover the lower part of the boles, and some trees have large, charred scars at their bases. Lumber from these trees is prized for its straight dense grain and few knots.

The mountainside forest looks like a park, with a luxuriant ground cover of grass, wildflowers, and low shrubs, but few small trees. In some parts of the forest, trees are 1 to 2 feet in diameter and 200 to 230 years old. In other places, larger trees are as much as 4 feet thick and 400 to 600 years old. The 200-year-old trees arose after an unusual, high-intensity fire swept up this steep mountainside in about 1663, probably driven by a strong westerly wind. Normally this forest burns in low-intensity surface fires at intervals of 15 to 30 years. These underburns kill most of the small, slow-growing saplings that develop after the last fire. The flames scorch and prune the lower branches of overstory trees. Although small numbers of mature Douglas-fir stand scattered around, their numerous saplings fail to develop into thickets because of the recurring fires.

Now it is 1919. The forest sees its last fire, an underburn. From now on fires will be successfully suppressed. Consequently a crop of little firs that becomes established in the 1920s gradually develops into a dense understory. Fast forward to the 1950s and 1960s. Logging extends up these steep slopes, and most of the pine and larch are cut except for a 15-acre patch. The numerous skid trails and burned slash piles make microsites where vigorous saplings of larch, pine, and fir spring up. It's not clear why loggers spare the 15-acre parcel, which consists mostly of the smaller old growth trees that had become established after the 1663 fire.

Thirty years later, in the 1990s, a forester takes note of this parcel of old growth. By now the trees are a few inches larger in diameter and about 300 years old. He suggests this grove be protected as old forest for biological purposes and as an educational area, since it is accessible and located near the city in the valley below. Next, the forester's attention turns to the dramatic changes that have taken place in this unlogged patch of forest as a result of preventing fires. By now, an understory thicket of Douglas-fir has developed and is already infected with dwarf mistletoe. Partly as a result of competition from this growth of fir, the old pine and larch trees are dying at an alarming rate. Also, the firs have developed into a dense layer of ladder fuels, putting the stand at risk of burning completely in a severe wildfire.

This is the story of an actual mountainside forest high above the city of Missoula in western Montana (Arno, Smith, and Krebs 1997). Here is how we unraveled the history of this pine–larch forest.

History in an Ancient Stump

The large black remnant of what had been a gigantic and very old ponderosa pine still stands on the steep forested mountainside. Despite its heavily charred surface, much of the stump is rock-solid, and beads of hardened pitch are scattered across its flat-sawn top. The stump shows that many fires had eaten away the uphill side of the tree. A black, triangular wound that foresters call a "catface" contains a record of the fires (Fig. 8.1). Vertical seams covered by folds of healing wood tissue line both sides of the catface. Each of 13 different folds on one side of the catface is a scar from a surface fire and the tree's subsequent growth response as it grew over the edge of the wound.

Figure 8.1. Catface at the base of a ponderosa pine shows evidence of at least two fire scars. The inverted V-shaped patch of wood killed by the first fire was later charred by the second fire and is black. The wound inflicted by the second fire lies above the char. The tree is slowly healing over, and without further fires will eventually envelop the catface. (Steven Allison-Bunnell photo.)

To learn the actual dates of the 13 fires, the stump must be carefully cross-sectioned. First, we scrape away the mound of duff and the mat of kinnikinnick from the base of this 3-foot-high charred relict. We examine the fire scar folds carefully to see at what level on the stump they are most intact and free of rot. Soon our chain saw's 36-inch bar labors through the hardened pitchwood. Within a few minutes the fresh-cut surface reveals the marvelously clear pattern of annual rings that records the tree's 600-year life.

The salvaged slab is an irregular band several inches wide and 4 inches thick that extends from one outer edge of the stump through the center to the opposite edge. One side of the originally circular tree had been eaten away by many fires, and the opposite side of the stump is rotting away. Only the pitch-filled wood near the catface survived. The 4-foot-long slab weighs about 200 pounds, and we struggle to carry it up the steep brushy slope to our truck.

Sanding the ancient remnant in the laboratory reveals even the smallest growth rings, which show the number of years (the interval) between each of its fire scars (Fig. 8.2). Counting and recounting confirms our suspicions. The seven most recent fire scars on this stump occurred at intervals coinciding precisely with the dates of seven consecutive fire scars identified on living trees in the adjacent old growth stand. These seven outermost scars were from fires in about 1889, 1869, 1849, 1815, 1798, 1777, and 1747. There were 30 annual rings between the most recent fire scar (ca. 1889) and the outermost ring on the stump. That placed the year this huge tree was logged as approximately 1919 and explains why it did not record the 1919 fire, as did the living old growth trees.

Most exciting is the sequence of six fires predating any fires recorded on the living old growth trees. The good condition of the catface on our remnant suggests that five of these early scars, dating from about 1715, 1686, 1663, 1628, and 1594, represent a relatively complete record of the fires that had swept across this mountainside. The tree's earliest remaining growth rings date to the fourteenth century, but the center (pith) had burned away along with any evidence of fires prior to about 1500. The earliest fire scar on the stump dates from about 1523; it appears that scars from other fires between then and 1594 could have been burned away.

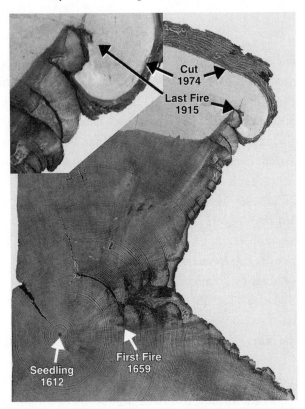

Figure 8.2. Cross section of a ponderosa pine with scars from 21 fires between 1659 and 1915. The inset shows the typical healing pattern of overlapping growth rings that tend to eventually cover the wound. Adjacent trees showed scars from additional fires that probably did not burn this tree to any extent because of very light fuels. (Arno 1976, USDA Forest Service photo.)

Most trees in the adjacent old growth stand are ponderosa pine and western larch that became established between about 1665 and 1690 (Arno, Smith, and Krebs 1997). It seems that this even-aged stand of pioneer species could only become established despite competition from Douglas-fir and grand fir if a high-intensity fire had killed most of the forest, leaving an open burned area. Here on the stump remnant was the evidence of that event, a fire scar from about 1663. The sequence of pre-1900 fires at average intervals of about 27 years extends back to 1594. This was 200 years before the first European fur traders and explorers penetrated this region.

Looking for Clues

You can apply this process for discovering forest history to the kind of wildland forest you are most likely to encounter in the West, which commonly sprang up after one or two past logging episodes. First look for a few widely scattered old growth trees greater than 200 years of age that predate historic logging. They typically stand out, being unusually large, having big, twisted branches and a craggy old appearance, or having bark distinctly different in color and texture compared with younger trees of the same species. The relict trees, remnants of the historical forest, may have been so crooked that loggers rejected them. You can watch, too, for dead old growth trees still standing (snags) or fallen. Live old trees provide the best and most easily interpreted records of past fires. However, such trees are now rare or absent in many forests. Old snags often display pre-1900 fire scars, but because of rot they can be dangerous to cross-section. Also, the resulting cross sections are heavily decayed, and the year the tree died, indicated by the outermost growth ring, is difficult to determine.

THE MERITS OF STUMPS

Stumps are frequently the best source of fire history information because they are relatively abundant and the stump top already provides a cross section to examine. Stumps can be sectioned easily with a saw and with negligible environmental impact.

Stumps from old trees may have three kinds of evidence of fire history. Growth rings on the stump top may show the curled pattern of a buried wound that had healed and become covered with bark by the time the tree was cut. You may find an open fire scar on one side of the stump. Old char on the stump's thick bark, but not on the cut surfaces, indicates a fire before the logging occurred. Even heavily decayed stumps can clearly show these features. However, only a few of the old stumps from a former forest are likely to show good evidence of low-intensity fires since the majority of old growth trees were not scarred because of their thick bark and light fuels. Also, many of the original stumps may have largely disintegrated. You must therefore search thoroughly to locate the best of this fragmentary evidence of past fires.

Stumps of trees logged with chain saws generally average about 10 to 20 inches tall and were cut after 1945. Low stumps like these with only about 100 annual rings or less will probably not have many scars from pre-1900 fires before the advent of suppression. However, if the trees are of a fire-dependent or pioneer species and were all about the same age when cut, the year when they germinated may be shortly after a forest fire.

Stumps of old growth trees (more than 200 growth rings) are a good source of information. These may be relatively recent, low stumps without extensive decay or high stumps 3 to 12 feet tall from logging in the late nineteenth or early twentieth century. The tallest stumps were left by fallers working atop "springboards," or platforms inserted into notches cut in the trunk several feet above the ground. This was done to avoid having to cut through the swollen and unusable bottom of the trunk, which often contained multiple fire scars and large amounts of pitch. Sometimes the big old trees were felled from a lower stump, but then the base of the fallen tree, containing fire scars, was sawn off and left in the woods. This huge block of tree trunk is called a "long butt." Long butts and tall stumps often reveal the number of historical fires, although they may be too rotten to section and date.

In many forests it is possible to discover the pre-1900 fire history, species composition, and general structure just from studying the stumps. You can often make a rough estimate of the age of the tree when it was cut by counting (or a combination of counting and estimating) the number of growth rings on some of the better-preserved stumps. If a stump is still sound, you may clearly see growth rings on a cross section sawn from it. If the stump is largely rotten, its weathered top often provides the best representation of the growth rings and may be most easily counted in place in the forest.

SINGLE FIRE SCARS

If your search of old growth trees, snags, and stumps has revealed apparent fire scars, they need to be evaluated. When you find only single scars, it is important to see if they show up on two or more stumps or old trees to confirm that this was a fire that spread through the forest and was not just confined to a small spot. Also be sure that fire caused the scar. Bears and other animals, bark beetle attacks, scraping damage from falling trees, and logging damage can also make scars at the base of trees.

If the stand is on sloping ground, look for fire scars on the uphill side of the trees. Fire scars are usually an inverted V, with the widest part at the ground line. Other kinds of scars often have an irregular shape. Bear scars often have shreds of loose inner bark hanging from the upper edges. Porcupine scars commonly appear as several patches, many not extending to the ground. Bark beetle scars, which appear in lodgepole pine, tend to be more irregular in shape than the inverted V of most fire scars. Scrape scars from falling trees are usually very tall and narrow, and the fallen tree is still present. Old, oval scars on ponderosa pines throughout the West were made by American Indians. They peeled the trees in spring to use the sap layer (cambium) as a sugar-rich food, but the bottom of these scars is usually 1 to 3 feet above the ground (Barrett 1985, Swetnam 1984). Logging scars often have splintered wood from the impact of mechanical damage, unlike fire scars that involve only lethal heating of the tree's cambium. Char on the adjacent bark confirms that fire was responsible for a single scar. Although this char eventually weathers off, it commonly can be found on Douglas-fir and some other thick-barked species for about a century after the fire. Distinguish char from black lichen, a primitive plant that looks like a roughened outer coating. Char will rub off, leaving black residue on your fingers. Rubbing lichen does not create black residue.

MULTIPLE FIRE SCARS

Some species, like the fast-growing coastal Douglas-fir, seldom have open fire scars (catfaces) because they rapidly heal over any wounds. You can best detect their fire history by examining stumps for buried scars. However, many species, including inland Douglas-fir and ponderosa pine do retain catfaces. That fire is the cause is confirmed by char on the exposed wood in the center of the wound. The first fire kills the cambium in the wound area, and over several years the charred bark falls away. This reveals a smooth light-colored wood surface, usually in an upside-down V. A second low-intensity fire is likely to char this exposed wood and kill a band of cambium bordering the original wound. This same process is repeated with each additional fire, so the most recent or outermost fire scar will not exhibit charred wood (Fig 8.1), although it may still be covered by dead, charred bark.

It takes a lot of practice to accurately interpret the number of fires in a complex or partially consumed catface. If you are lucky, you will find a catface that is more or less symmetrical, consisting of two or more inverted Vs nested one inside of another. However, this idealized pattern may not appear for several reasons, such as consumption of part of the catface during one of the fires. In a well-formed catface, you can estimate the total number of fires by counting the maximum number of individual fire scars at any point on either side. Each individual fire scar is represented by a vertical seam accompanied by a fold of healing tissue.

It is important to distinguish vertical cracks and narrow seams caused by weathering of the exposed wood in a catface. Do not count these as individual fire scars. Inland Douglas-fir often heals wounds with irregular growth patterns that you can mistake for numerous minute fire scars, especially if you only evaluate a cross section and you don't look at scar patterns within the catface. Actual fire scars are followed by consecutive healing growth rings that consistently overlap and envelop each other, slowly growing over the wound (Fig. 8.2). Because of the scarcity of well-preserved and complete fire scar sequences in most forests, it is important to search thoroughly for such evidence. To ensure your findings are reliable, it is desirable to locate repeated, similar patterns of the maximum number of fire scars, and to cross-section a few of the best preserved of these in a given forest stand.

SAMPLING STUMPS

Determining the date of a fire scar on a living tree using an increment borer is a difficult art that applies best to a single scar (Barrett and Arno 1988). You can also age the live trees in the stand using an increment borer. (To buy an increment borer, contact your local forestry agency for the address of a forestry supply company.) Dating fire scars on a living tree using a chain saw requires sawing skill (be sure to follow the manufacturer's safety recommendations) and may damage trees (Agee 1993, Arno and Sneck 1977, McBride 1983). It is simpler to concentrate on fire scar patterns on stump tops or cross sections sawn from stumps as already described.

You can quickly profile the fire history in a stand or small area by finding the individual stumps or old trees with the largest number of fire scars. To ensure your profile's validity, confirm that a similar number of fire scars is found on some other stumps or trees. If fires were quite frequent, any individual chronology probably underestimates the number of fires that occurred in a given period (Arno and Petersen 1983). You can simplify your investigation of fire history by thoroughly inspecting the fire scars in the stand or small area, evaluating the soundness of the stumps that have the most complete sequence of fire scars, cross-sectioning the least decayed or damaged of these stumps, and sanding and analyzing only samples that are of good quality. You can waste a lot of time studying heavily decayed or damaged sections that will not yield conclusive information.

When working with very old stumps, it is often easier to determine the number of years in each interval between fire scars instead of the date of the latest fire scar. Even if you cannot date the fire scars on a stump with any confidence, knowing the *intervals* between fires is useful for identifying the historical fire regime.

You may be able to approximately date all fire scars and the tree's year of origin by simply counting growth rings if you can determine the year of logging (the outermost ring on the stump) or the year of the most recent fire scar. You may be able to find historical information—for instance old fire maps at a national forest ranger station or old newspaper accounts kept in a local library or by a historical society—to date the early logging or the most recent fire. You might date the logging by increment boring remaining old trees whose neighbors were logged. The surviving trees often show a sudden, dramatic increase in growth rate when the competing trees were removed.

The sapwood between the latest scar and the year the stump was cut may be partly missing or badly decayed and may have very fine, indistinct annual rings. Sometimes you can count to the outermost ring at some point on the stump's perimeter. You will get the best count if you sand the cross section smooth or slice the fine rings with a razor blade, wet with water, and use a 10-power hand lens to count the growth rings (Arno and Sneck 1977). If it is not possible to take a stump cross section home for examination, you can study the intact stump in the woods.

Fire-Initiated Age Classes of Trees

When a fire heavily thins or kills a forest, a flush of new tree regeneration is likely to follow. You can identify such a cohort of same-aged trees as a fire-initiated age class even if it was first established centuries ago. These age classes of living trees are a testimony to fires of the past. They are most evident in mixed and stand-replacement fire regimes, where fires were infrequent and created room for new trees. The pioneer tree species require open conditions to regenerate in large numbers. Thus, "even-aged" stands of these species (including lodgepole pine, western larch, and coastal Douglas-fir) occur after major disturbances like fire, severe blow-downs, insect epidemics, or heavy logging. The new trees initially grow fast on the open site and decades later will remain similar in age, size, and appearance. Rocky Mountain lodgepole pine is famous for establishing soon after a major fire and producing dense, even-aged stands of small trees. You can confirm that fire initiated the new stand by finding charred remnants from the former forest and fire scars on surviving trees at the edge of the stand.

Approximate the year of the fire by counting the number of annual rings near the ground line of several trees within the apparent age class. Take increment borings from trees not far above the ground line, and count only borings that hit or nearly hit the pith. You need a minimum of five trees of similar ages to identify a fire-initiated age class. Ring counts from new stumps are an alternative to increment borings. If there is quite a disparity among the ages, try taking additional samples to help clarify the situation. The oldest individual trees within a single age class will best indicate the year of the fire. For example, if by late summer 2000 the trees sampled on an old burn had total ages, respectively, of 88, 87, 85, 82, and 80 years, the fire probably occurred in 1912 or a few years before. If historical records indicate fires burned in the general area in 1910, that would likely be the actual fire year.

Sometimes the most recent fire in an area leaves a few large surviving trees of a pioneer species, which might date from a previous fire. The same age-class dating process can be used on these survivors to estimate the year of the earlier fire. If fire killed the former stand several decades ago, but the old dead trees are not badly rotten, you can determine the date of the previous fire from sections sawn from the base of the fire-killed trees.

Reconstructing Old Forests

The dates and intervals of historical fires are just one part of the picture. To complete our portrait of a forest before fire suppression, we need to know more about the trees that grew there. Forest ecologists painstakingly sample and analyze their results to make a detailed reconstruction of how a forest appeared prior to fire suppression (Arno, Scott, and Hartwell 1995, Bonnicksen and Stone 1981, Habeck 1994, Stokes and Smiley 1968, White 1985). However, it is much easier to make a general characterization of the historical forest provided that it has been logged only once.

You can apply the following approach to the particular place you want to understand. Many of the stands you are likely to study were logged roughly 100 years ago. If the stand was logged more recently, you can adjust this method to fit that situation. To reconstruct the stand, you want to describe its structure just prior to logging, which for our example occurred in 1900. Stumps from early logging will still be identifiable in your stand and will provide you with one type of information. A second category of information is remaining "old trees," or trees that were here in 1900 and were not cut. The third category is "young trees" that were not here in 1900.

After examining the general structure and variation within the stand, choose an area having average conditions and lay out a quarter-acre (105-foot-square) plot. Within this plot, make a thorough search and list all the old stumps by species. For species with thick bark, remnants of bark usually remain near the ground line. You can slice them with a knife and compare them with sliced mature bark on live trees to identify the species of the stump. You can also use common characteristics of the wood of local tree species for field identification. These include the pitchwood in old ponderosa pine stumps; aromatic and easily whittled western red cedar or incense cedar *(Calocedrus decurrens)*; light-colored, nongrainy spruce; and wood of old western larch that commonly separates and falls apart along the rings.

Next, determine the approximate ages of the largest trees. You will need to tally all trees greater than 100 years old by species in the plot. These can be tentatively identified based on size (diameter at 4.5 feet above the ground) and appearance (such as plated orange bark in old ponderosa pine). If there are few old trees, you can determine the age of each one by increment boring. If there are many old trees, you can identify ages of several that represent the different diameters and species.

With this information you can characterize the historical forest (ca. 1900) using a simple table. The table lists the number of old trees in the plot (or per acre) by species, which is the total of the old stumps, snags, old fallen trees, and surviving old trees you tallied. It is helpful to estimate how many of these were apparently overstory trees in the historical stand (surviving trees older than 170 years in 2000) and to estimate how many were small trees in the original stand. You can then compare the historical stand with the current stand. You will see how the number of overstory trees per acre by species in the historical stand differs from the modern stand. Table 8.1 gives data of this sort from an actual stand. This example shows a steep increase in small trees, especially Douglas-fir, since 1900. The data for historical stands tend to underestimate the number of trees, especially small trees, because some would have died and decomposed and not been counted in the tally. Large trees that died can be identified and

Table 8.1. Number of trees per acre in 1900 and in 1991 in a dry-site forest in western Montana. No logging has occurred, and the last of relatively frequent fires was in 1889. PP = ponderosa pine; DF = inland Douglas-fir (Stand B-2 Arno, Scott, and Hartwell 1995).

	1900		1991	
	PP	*DF*	*PP*	*DF*
Small trees (<100 yr)	6	19	122	1,010
Large trees (>100 yr)	17	13	24	25

should be added to the data for the historical stand. In most cases the portrait of the historical stand should still be reasonably accurate.

This evidence will usually allow you to identify the historical fire regime as understory, mixed, or stand-replacement. For example, if the historical stand was well stocked with trees with fire scars that record several fires, this indicates an understory fire regime. Conversely, if the historical stand was even-aged and did not experience or survive a fire, this suggests a stand-replacement fire regime. Mixed regimes will not fit these scenarios and are likely to be represented by a historical stand that had some fire-surviving trees along with many younger trees.

Practical Use of Fire History

If the life of the forest since the last ice age had been captured on motion picture film, today would represent only the last frame. In many western forests where fires were infrequent, modern conditions may still be similar to some of the historical conditions shown in the movie. However, in forests where fires burned frequently, conditions like those found today may never have existed over thousands of years prior to logging and fire suppression. Studying the history of any particular forest and its fires can tell us if and how it has changed from forests of the past.

Even just observing the historical species composition and the associated fire regime, without gathering data, can be helpful. In a given place, these observations create one snapshot of forests in the past that were produced under the influence of natural fires. Similar studies from many locations in a given forest type collectively provide many snapshots, stronger evidence of historical conditions. This is like archaeological work, reconstructing past tree communities from surviving evidence. Some sites have better indications of past forest structure and disturbance patterns than others. Good descriptions of past forests require careful sleuthing and scrutiny of what remains. Sometimes, as in areas that were heavily logged and the slash piled with a bulldozer and burned, hardly any trace of the former forest remains.

Reconstructing a former natural forest can give you insight into the historical disturbance processes in a given location and the kind of forest that developed as a result. Connecting the dots of cause and effect in the natural environment will give you a better understanding of forest ecology. This can allow you to evaluate or to help plan ecologically based management for a forest you are concerned about (Swetnam, Allen, and Betancourt 1999).

Chapter 9

Fire-Prone Forests:
Can We Adapt to Them?

In August 1994, the Star Gulch wildfire roared through 30,000 acres of rugged terrain on the Boise National Forest near Idaho City, killing most of the trees (Barbouletos, Morelan, and Carroll 1998). The following spring, this burn revealed a striking dichotomy. The forest on one side of a road was completely black and stripped of foliage to the top of towering 140-foot ponderosa pines. On the other side, the trees next to the road had scorched orange-brown needles, but behind them, although the ground was blackened, the entire forest was still green.

What was the difference? The whole area was a heavily stocked forest of large ponderosa pines. The terrain and forest structure were similar, and the forest had never been logged. Yet on one side of the road the fire had been an inferno, while the other side experienced only a low-intensity underburn. The underburned stand survived because it was part of a 1,000-acre block that had been prescription burned in the early 1980s and again the spring before the wildfire. The severely burned forest had not seen fire of any kind in many decades. When the raging Star Gulch wildfire entered the previously underburned area, the sparse surface and ladder fuels no longer supported canopy burning. This transformed the crown fire to a low-intensity underburn. When the wildfire emerged from the other side of the previously treated block, it resumed high-intensity burning. Classic fire suppression tactics, including building firelines to contain a fire, have trained us to think that fires respect physical features such as roads or streams. But as this fire shows, the road was not a boundary at all. Rather, the fire history of the stand itself created an essentially biological boundary that dramatically changed the fire's behavior (Barbouletos, Morelan, and Carroll 1998).

What happened on the Star Gulch fire is just one of many incidents in the western United States in which a severe wildfire calmed down when it entered a stand where the fuels had been treated. Of course, the effectiveness of different fuel reduction treatments depends on many factors, and treatments will not prevent stand-replacement burning under all conditions. But they are frequently far more effective than any heroic measures on the fireline. Although recent experimental burning and fire modeling have scientifically established the relationship between fuel reduction and diminished fire behavior, it is not really a new discovery (Agee and others 2000, Anderson and Brown 1988, Keane, Arno, and Brown 1990, Martin, Kauffman, and Landsberg 1989, Pollet and Omi 1999, Scott 1998, Stephens 1998). Within the forestry profession and the U.S. Forest Service, the idea that we can adapt to fire-prone forests by managing fuels struggled alongside the more orthodox advocacy of all-out fire suppression almost from the beginning. In the last 20 years this notion has percolated through the forestry profession and even been embraced by some government leaders. What remains is considerable public confusion and ambivalence over the proposition of thinning and purposefully burning hundreds of thousands of acres of public lands. We want to make the case that an appropriate fuel management strategy for every landscape zone, from residential forests to wilderness areas, is not only ecologically and biologically desirable, but is the only way to prevent the catastrophic fires that concern suppression advocates and conservationists alike.

Prophets of Prescribed Fire

Recent research demonstrating that wildfires lessen in intensity when they enter lightly fueled stands confirms the recommendations of John Wesley Powell and the pioneer timbermen who advocated light burning to reduce wildfire hazard (Hoxie 1910, Powell 1891, Pyne 1982, 2001a).

But it has taken the better part of a century for this insight to regain credibility. During the 1920s, Stuart Show and Edward Kotok of the U.S. Forest Service studied fire scars on stumps in pine forests of northern California and concluded that frequent fires had been characteristic of the area (Show and Kotok 1924). They concluded that the fires had damaged some overstory trees by creating scars. While this agrees with current findings, they also judged that by killing saplings, fires had kept forests unnecessarily open. Not accounting for the semiarid climate, Show and Kotok believed that more trees could and should grow in these open stands. To them, even low-intensity understory fires reduced the forest's potential to produce timber and threatened the forest resource (Pyne 1982). Valuing the quantity of trees over any other measure of forest health or productivity prevailed in forestry until the 1960s, and the Forest Service strongly opposed using fire in western forests other than for disposing of piles of slash (Pyne 1982, Schiff 1962).

However, in 1943, Harold Weaver became the first professional forester to argue the case for using fire to manage western forests. Weaver had grown up in the ponderosa pine country of Oregon's Blue Mountains. He earned his forestry degree at Oregon State University in 1928, where he was "thoroughly imbued . . . with the incompatibility of pine forestry and fire" (Weaver 1968, 128). As a young forester on the Klamath Indian Reservation in southern Oregon, Weaver met several experienced woodsmen who considered the policy of excluding fire from the forest a serious mistake. Skeptical, he asked how the forest could regenerate if frequent light burning were practiced. One veteran logger responded that fires in the past were quite low intensity and just crept through the pine needles, so that many young pines survived. In the 1930s, Weaver met the renowned forest entomologist F. P. Keen and was "shocked" to learn that fire had an important ecological role in maintaining ponderosa pine forests. Keen showed Weaver the multiple fire scars on both old and young pines that proved they had survived low-intensity fires (Weaver 1968, 128).

As time went on, Weaver examined ponderosa pine trees and stumps throughout much of the West. Everywhere he found unmistakable evidence that these forests had experienced frequent low-intensity fires in

the past. He concluded that such fires are part of the natural environment
and that they were an important force shaping the magnificent open ponderosa pine forests that historically characterized much of the West. As a
result of this conclusion, Weaver began experimenting in the late 1930s
with prescribed burning to reduce wildfire hazard and thin excessive
numbers of saplings in ponderosa pine forests.

Weaver's first treatise on the need for prescribed fire in forestry and the
ecological damage caused by excluding fire must have seemed rank heresy
when it appeared in the *Journal of Forestry* (Weaver 1943). Following his
article, the journal printed an anti-fire rebuttal by A. A. Brown, who later
became the director of fire research in the Forest Service (Brown 1943).
Over the years Weaver's many articles on this topic were required by his
employer to carry the disclaimer that these were "the author's views only
and . . . not to be regarded in any way as an expression of the attitude of
the Indian Service [later the Bureau of Indian Affairs] on the subject"
(Weaver 1943, 7).

Surprisingly, Weaver's radical advocacy of fire in forestry and Brown's
critique evoked only one letter of response to the *Journal of Forestry*. A
retired forester, Samuel Lamb, wrote to praise Weaver's proposition:

> Arthur A. Brown, in his comments on Mr. Weaver's article,
> says we must substitute harvesting by logging for nature's
> harvesting by fire, . . . but Mr. Weaver is not talking about
> "harvest" by fire, he is discussing fire as a cultural tool. Now
> that we know the cost of widespread thinning by use of the
> axe, let's see if we cannot find a cheaper way through the use
> of controlled fire. Let's pry the lid off the fire question, stop
> crucifying all "pyromaniacs" among us, and give controlled
> fire a real tryout in all the forests where it can be useful
> (Lamb 1943, 295).

Lamb's distinction between letting fire burn trees instead of cutting
them down and using it to improve forest health and growth remains mis-
understood to this day.

In the 1940s and 1950s Weaver convinced some Forest Service
researchers to join him in studying the effects of prescribed burning
(Weaver 1968). In 1949, Russell LeBarron, a Forest Service scientist who

had been in touch with Weaver, encouraged consideration of various uses of fire in forest management. The *Journal of Forestry* later published LeBarron's views because the editor felt he offered "an unusually objective analysis of this controversial subject" (LeBarron 1957, 627).

During the same period, Harold Biswell of the University of California's School of Forestry also began experimenting with prescribed burning (Biswell 1989). Biswell had been involved in prescribed burning in southern forests and felt it should be tried in the West also. Over more than three decades, Biswell introduced large numbers of students to fire ecology and prescribed burning. He also mentored a generation of fire ecologists and fire managers who taught at colleges and universities and worked for land management agencies throughout the West.

Weaver and Biswell found that in addition to fuel reduction, prescribed burning offered other benefits. These included maintaining forests in a more open condition, which allowed trees and saplings to be more vigorous and improved forage for livestock and wildlife. Weaver and Biswell gave many presentations on the benefits of forest burning at forestry conferences and field workshops beginning in the 1950s. By the late 1970s, Weaver, Biswell, and many other prescribed burning practitioners from all over the West had gained wide acceptance for using fire to maintain ponderosa pine and related forests (Biswell 1989, Kilgore 1976, Nelson 1979).

Forestry Accepts Fire, but Other Barriers Emerge

Ironically, by the time broader support for using fire finally emerged within the forestry profession and the U.S. Forest Service, there were new barriers to applying prescribed fire. First, many fire-dependent forests had deteriorated, making it harder to restore them. The larger, vigorous trees had often been logged. This selective logging and many decades of fire suppression often fostered dense thickets of small trees. Many stands were stagnating due to intense competition, which allowed insect epidemics and diseases like dwarf mistletoe and western budworm to build up. Surface and ladder fuels had accumulated so that it was now difficult to burn safely or effectively without first removing many small trees. Under these conditions an effective thinning fire might badly stress the surviving trees and create too much new fuel by killing trees. Conversely, a low-intensity fire might accomplish negligible thinning or fuel reduction. A second new barrier to prescribed fire was the formidable array of laws and regulations now protecting certain features of the environment such as threatened species, air, or water quality, but with no concern for restoring the natural environment itself by returning fire to the forest. Third, perhaps a million new homes and cabins had been built in western forests, many at great risk to wildfires. Finally, millions of environmentally concerned citizens were now rightfully involved in shaping policy and actions on public forestlands. Most had absorbed the forestry profession's previous message that fire in the forest is inherently frightening, ugly, and dangerous to all forms of life. Most of these concerned citizens had not been informed of the vital role that fire plays in maintaining wildland ecosystems. This new message is not only more sophisticated and complex than the old one, but its urgency is even greater.

The very idea of stewardship means taking good care of what is entrusted to us. Simply leaving today's forests alone after a century of fire suppression and forestry focused on extraction of big trees is not caring for them; it is abandonment. Wildlife biologists know that an animal raised in captivity must be trained to survive before it can be released. In a sense, we cannot expect forests that we have attempted to domesticate for a long period to "return to the wild" unaided. Good stewardship of wildland forests in the West begins with the fact that the natural fire regime in each area produced vegetation and wildlife well adapted to a fire environment. Any radical alterations we make in this pattern of natural disturbances or in the kind of vegetation that it produced put the forest ecosystem at risk (Smith 2000b). And one of our most radical alterations has been fire exclusion itself. Attempting to remove fire from forests that in the past burned frequently has brought on epidemics of insects and diseases and wildfires of unprecedented severity (Covington and others 1994, GAO 1999, Mutch and others 1993).

Not only have foresters had to change their attitudes toward fire, ecologists have had to change their picture of ecosystems existing in static equilibrium. Ecologists now see that disturbance of all scales, from the toppling of a large tree to a hurricane that flattens a swath of forest, in all environments, from marine to terrestrial and tropical to temperate, is a driving force in creating and maintaining biological diversity (Botkin 1990, Foster, Knight, and Franklin 1998, Knight 1987). The steady-state "climax" community treasured so long in ecology is now just one temporary end point in a system that is incomprehensibly dynamic (Daubenmire 1968). Today we see that the stability within a naturally functioning forest ecosystem is not a perpetuation of the species composition or age of individual trees on a single patch of ground, but the overall array of different stand structures, successional stages, and plant communities on the totality of patches across the landscape. Fire history studies of the Yellowstone ecosystem illustrate this. From the late seventeenth century to the early twentieth century, the landscape of Yellowstone National Park was like an enormous patchwork quilt made of stands of different ages created by occasional stand-replacement fires (Knight 1991, Romme and Despain 1989). Then, with fire suppression gaining effectiveness in the early decades of the twentieth century, the Yellowstone landscape slowly began to lose its diversity as more of the patches advanced in age and an ever-increasing area of old forest emerged. This process of widespread advance of stand ages across the forest landscape, followed by unusually large and severe fires that replace forests on an unusually broad scale, is more dramatic in areas that historically experienced shorter fire intervals such as in the mixed fire regime. In Glacier National Park's North Fork Flathead Valley, the period of effective fire suppression equaled or exceeded historical fire intervals; thus the landscape mosaic aged, contributing to a very large severe fire in 1988 (Barrett, Arno, and Key 1991).

In many ecosystems it is neither possible nor desirable to create or preserve an unchanging condition, such as trying to protect old forest from natural disturbances. Moreover, the pursuit of such a goal can damage the ecosystem (Botkin 1990). In contrast, the way to perpetuate disturbance-dependent ecosystems is to maintain a semblance of natural disturbance processes interacting with the native plants and animals. In western forests, disturbance and change are inevitable and necessary. Like it or not, many or even most of those changes will be human-caused. For nearly 100 years forestry has operated under the rationale that logging substitutes for fire by removing trees that fire would otherwise "waste" stone Ecosystem. A single patch of old growth trees *is* precious when there are none left around it. We do not mean to be cavalier about efforts to protect these places and their unique plant and animal communities. We see restoration forestry as a way to create more old growth rather than whittling away at it or allowing it to be predisposed to severe wildfires or competitive stress and epidemics of insects and diseases. Without restoration forestry to expand the pool of functioning patches of habitat, the remaining relicts will eventually die, their like never to be seen again.

A key element to managing fuels and fire for ecological restoration is to create forested fuel breaks in strategic locations. These are open stands of tall trees with only light fuels on the ground (Agee and others 2000). Broad bands of forest having sparse fuels are safer zones for crews attempting to control a wildfire. Conversely, roads or narrow firebreaks provide little protection from high-intensity flames, and wildfires easily jump them. A strategically situated network of forested fuel breaks could be created by treating a small percentage of the forest specifically selected to reduce wildfire hazard to homes and other developments. The treated areas, mostly accessible by road, are places where there is a good opportunity to control the advance of a wildfire. Also, the presence of these treated zones would allow us to make broader use of prescribed fire and natural fires to restore forests in natural areas that lie beyond the forested fuel breaks. Management of fuels and fire can be applied to the forest landscape using three different strategies linked to the three types of land management zones: the residential forest, the general forest, and wilderness and other natural areas.

THE RESIDENTIAL FOREST

The "residential forest" surrounds homes, summer cabins, resorts, and communities. Dead and living forest fuels normally accumulate in these forests, and many of the landowners are unaware of the associated hazards or what they could do to reduce them. Buildings in this zone often compound the wildfire hazard by being highly flammable themselves and having flammable materials connected to them. When fire threatens the residential forest, attempts to control the fire's spread are compromised or largely abandoned because suppression resources and priorities are shifted to protecting homes and other developments. Even if we want the residential forest to look and feel "natural," it must be tended frequently to control hazardous fuels. In fact, forest residents don't have to sacrifice aesthetics to reduce the fire hazard of the surrounding forest (Schmidt and Wakimoto 1988, Scott 1998). Ultimately, if you live or own property in the residential forest, you should consider the cost and effort of fuels management to be just as much a part of the routine maintenance of your property as mowing the lawn or repairing a leaking toilet. Homeowners who allow hazardous fuels to accumulate on their property must be prepared to assume the risks of uncontrollable fire and the cost of damage to their property and adjacent properties.

The fire that burned through Los Alamos, New Mexico, in May 2000 tragically illustrates the consequences of failing to create a forested fuel break. Although the Cerro Grande wildfire burned as a high-intensity crown fire in certain areas, the flaming front spread through the town of Los Alamos (population 11,500) as a low-intensity surface fire (Cohen 2000). Despite its low intensity, the fire destroyed over 200 residences in this community located within a ponderosa pine forest. As fire approached Los Alamos, it was burning in pine needle litter and other surface fuels, seldom killing trees. This was a large fire, producing great quantities of smoke, and as it entered the residential area, the fire spread through pine needle litter directly to homes that had combustible material such as flammable landscaping, wooden decks, woodpiles, and shake roofs on and around them. When the buildings caught fire, the adjacent trees were often killed, but trees away from burning buildings generally survived the fire.

Some homes caught fire even though the entire width of a street separated them from other burned vegetation and buildings (Cohen 2000). These homes apparently were ignited by airborne firebrands, mostly originating from intensely burning homes and other buildings, rather than from the moderately burning forest fuels. In several cases a hand-raked fireline that removed pine needles from the base of a wood exterior wall kept the house from igniting. Although firefighters were able to save many homes, the high ignitability of most of the residential area allowed so many simultaneous house fires that suppression forces were quickly overwhelmed (Cohen 2000). One Los Alamos resident had recently thinned and removed fuels on his 3-acre homesite based on suggestions he solicited from foresters. He is now convinced that these treatments saved his home (Meeks 2001).

Fire damage to Los Alamos National Laboratory (LANL) facilities was limited to 28 outlying structures of minor importance (Webb 2000). Had the same fire occurred four years earlier, before thinning and fuel removal in adjacent forests had begun, many of the LANL facilities probably would have been destroyed. All of this evidence strongly suggests that if a forested fuel break had been completed around the entire Los Alamos area, and if residents had cleared combustible material adjacent to their houses, there would have been considerably less fire damage.

Even in forests that naturally burn in stand-replacing fires, forested fuel breaks could protect facilities and private lands. The controversy and high cost of fighting the 1988 fires in and around Yellowstone National Park could have been minimized if forested fuel breaks and measures to reduce building flammability had been in place. Government agencies spent about $145 million attempting to control these fires, but except for protecting structures, this massive suppression effort was ineffective (Wakimoto 1989). The resort complexes in Yellowstone Park and the adjacent communities of West Yellowstone, Silver Gate, and Cooke City were bordered by dense forests often containing accumulations of dead trees. Clearing out dead trees and small live trees from forests around these communities could have substantially reduced wildfire hazard. Some thinning was attempted as the flames advanced, but there was too little time to remove enough fuel to make a difference.

Television news covered the dramatic rescue of the towering log lodge at Old Faithful. The steep roof of this venerable structure clad with wood shingles was protected from firebrands at great risk by two firefighters standing on it with hoses while an inky black storm of smoke and fire engulfed the area. When the fire front reached the Old Faithful resort complex, large asphalt parking lots arrested the surface fire, but the lodge's shingle roof was like a mountain of kindling awaiting the rain of firebrands lofted over the entire area. In hindsight, application of a fire-proofing treatment or installation of a permanent rooftop sprinkler system might also have been able to save the building without putting fire-fighters' lives at risk.

Throughout that interminable 1988 wildfire season, ominous wind-storms tracked across the northwestern states once a week, fanning fires into conflagrations and forcing firefighters to retreat. As these dry cold fronts passed, several raging wildfires threatened or engulfed woodland homes. These wind-driven fires remained uncontrollable until the weather moderated. One exception was the Little Rock Creek Fire, which blew out of a rugged canyon in the Bitterroot National Forest, crowned through the forest, and headed toward rural homes northwest of Darby, Montana. Few firefighters, engines, or other resources were still available since most had been sent to the big fires in Yellowstone and elsewhere. It seemed probable that the small suppression force assembled to control the Little Rock Creek Fire would be forced to retreat. However, soon after firefighters arrived, the flaming front reached a large swath of ponderosa pine–fir forest that had had a variety of fuel treatments, including thinning and selective harvesting with slash burning, and a stand that had been underburned in a prescribed fire several years before. When the wildfire entered this fuel-treated area, burning intensity dropped off, allowing fire crews to control it (McKee 2000).

Today, a number of national forests, national parks, and community groups in the West are interested in establishing forested fuel breaks. Forested fuel breaks are open stands of trees whose crowns do not touch each other and are pruned so as to separate the lower crown from surface or ladder fuels. Small trees are scattered and do not form thickets, and slash and dead brush are kept to a minimum. In studies that rate the public's perception of scenic value, these parklike stands commonly score higher than dense untreated forests (Scott 1998). Still, the majority of homeowners and others who have a stake in protecting the residential forest have shown negligible knowledge or interest in such safety zones. Some actually oppose such treatments, contending that any thinning will actually look like industrial logging or it will open the area up to more use by motorcycles and all-terrain vehicles. Hundreds of homes have been destroyed and families displaced during some recent wildfire seasons. Considering the high economic and social values at risk in the residential forest, intensive forestry to control fuels and maintain the visual appearance seems like an attractive option.

When properly designed, fuel reduction in residential forests can often nearly pay for itself with the income from small logs, posts and poles, firewood, pulpwood, or wood chips for generation of electricity (Fiedler and others 1999, 2001, Scott 1998). Even where the treatment produces no commercial products, prescribed burning can be cost-effective when compared to costs of suppressing a wildfire. The cost of prescribed burning to reduce fuels often varies from about $30 to $400 per acre, and it is designed to avoid damaging overstory trees. In contrast, suppression of a wildfire in the residential forest zone commonly costs over $1,000 per acre, and the fire often kills most of the trees (GAO 1999).

THE GENERAL FOREST

"General forests" have road systems and are managed to provide wildlife habitat and water quality as well as a broad range of recreational activities, firewood gathering, timber production, and livestock grazing. Traditionally, selective logging has often meant, "Take the best and leave the rest." The largest, most desirable trees have been cut, and thickets or understories of young, unmerchantable trees are left behind to continue to develop into hazardous ladder fuels. Stands with low economic value cannot be treated effectively using commercial harvesting; consequently, they continue to accumulate fuels.

Clearcutting was designed to counteract some of the problems with selective cutting. Properly conducted, clearcutting removed all the trees and burned the slash or piled it with a bulldozer. Then a plantation of vigorous new trees was established and managed for another clearcutting after a relatively short rotation of between 40 and 100 years, depending on site productivity. Traditional clearcutting controls fuels, but its application on public lands is now limited by major ecological and visual concerns. From an ecological viewpoint, a modified form of clearcutting called variable retention harvesting is more appropriate for stand-replacement fire regimes in the general forest (Franklin and others 1999). Variable retention forestry requires leaving and perhaps burning patches of live trees and snags within the harvest unit to retain important structural features of the natural fire regime, making the unit's shape more irregular, reducing unit area, and planning on a long rotation comparable to the historical fire regime.

Proper timber harvesting can help optimize forest density and reduce fuels, including slash. The belief that any harvesting, thinning, or other fuel treatment will damage scenic values leaves heavy fuels in place, even though few people find the immediate aftermath of a severe wildfire to be scenic. Proper management of the general forest focuses on "leaving the best and taking the rest," and being very careful not to damage remaining trees or the soil. The goal is to favor healthy, long-lived trees of the historically dominant species and to remove excessive numbers of trees, including accumulations of small trees. Each time the stand is tended, slash, surface fuels, and ladder fuels are reduced. Some trees are allowed to live indefinitely and become snags. The stand may be reentered every few decades to maintain these desirable conditions, but this will often allow commercial harvest as a byproduct of doing the ecologically based treatment.

In past decades, forest management on public lands focused on efficiently extracting valuable timber. Today, public opinion has measurably shifted to the idea that public forests should be maintained largely for wildlife habitat, scenery, natural diversity, and recreation. For instance, a Gallup poll found that 68 percent of Americans favor protecting the environment even at the risk of curbing economic gain, while only 24 percent favored giving economic growth first priority (Gallup 1999). Also, a poll of private forestland owners in the West found that their highest priority is stewardship and restoration of their forest, and making a profit is less important (Hutcherson 1991). In principle, there should now be greater public support for shifting funding from fire suppression to maintaining desirable levels of fuels and to regaining at least some of the benefits of prescribed fire. Perhaps the main reason why this funding shift hasn't happened is that the old fire suppression mentality has become ingrained in our culture and institutionalized as a richly funded program connecting all levels of government and providing income for large numbers of manufacturers and contractors (The Economist 2001, Mutch 2001, Pyne 1982). Even without a complete change in funding priorities, prescribed burning averaged about half a million acres annually in western national forests during the late 1990s, and the Forest Service hopes to increase its fuel reduction program to about 3 million acres a year within a few years (Hilbruner 2000). Most burning is done in ponderosa pine–fir forests, but is of such low intensity that fire kills only a fraction of the smallest trees, thus failing to meet the fundamental objective of thinning overdense forests.

Forests that were historically in understory or mixed fire regimes have now missed two or more natural fire and thinning cycles. Because of this, they often need to be thinned and have the larger nonmerchantable and smaller merchantable trees removed to introduce prescribed fires that will mimic the historical fires. This sometimes requires felling along with piling and burning of slash or chipping slash for use as fuel in industrial boilers or for generating electric power. After thinning and removal, prescribed fire can be effectively applied to help maintain low to moderate wildfire hazard into the future (Covington and Moore 1994b, GAO 1999, Harrington 2000). Fuel treatments can normally be carried out utilizing existing roads, and sometimes aided by low-impact, off-road harvesting machinery such as track-mounted harvesters that delimb trees and cut them into logs, log forwarders, or helicopters.

"Stewardship contracts" are a novel mechanism to fund fuel reduction treatments on national forests and some other federal lands that focus on leaving the best trees and creating the most desirable habitat conditions (Little 2000, Stewart 1998). Stewardship contracts contrast with traditional timber sale contracts, which rely on the value of timber extracted from the forest to pay for the entire treatment. So far, stewardship contracting is developing slowly and tentatively because of barriers to revising federal contracting procedures, scant funding and personnel to prepare new contracts, and opposition from some environmental interests who object to any tree cutting and especially to utilization of trees for forest products (Little 2000, Moore 1999).

Unlike managers of public forests, private individuals who own forest-land are free to choose treatments that are best for the forest ecosystem in the long run, regardless of economic costs. In reality, most private landowners, large and small, are attracted to treatments that provide the greatest monetary return (Jay, Arno, and Arno 1998). Conservation easements are increasingly used to purchase development rights from owners of forestland, preventing future development. For example, a $12 million deal made in 2001 used money from several agencies to acquire development rights to 53,000 acres of Plum Creek Timber Company forestland in valleys of northwestern Montana (Devlin 2001b). In return, Plum Creek is allowed to continue to manage the land for timber production and must continue to provide public access for outdoor recreation. Conservation easements could go one step further and use their financial incentives to induce private landowners to practice better forest stewardship. The easement can set up a stewardship management plan specifying how ecological and fuel management goals will be achieved (Fiedler and Arno 2000, Reid 1998).

WILDERNESS AND NATURAL AREAS

Managers of wilderness areas, national parks, and some other natural areas may allow lightning fires to burn under previously approved criteria for weather conditions, fuel moisture, time of year, and so forth to maintain the natural ecosystems. In the largest wilderness and natural areas, these fires can help maintain or re-create a mosaic of old and young forests similar to what occurred prior to fire suppression. Often, however, lightning fires must be extinguished because they might eventually threaten lands and forest homesites outside the natural area (Brown and others 1994). As of 1998, only 88 of 596 designated wilderness areas in the United States, excluding Alaska, had approved fire plans that allow some natural ignitions to burn, and even these approved areas continue to suppress many natural ignitions (Parsons 2000). As a result, no wilderness area has fully restored fire to its historical role.

Construction of forested fuel breaks near the residential forest zone could help wilderness managers to allow more natural fires. Still, lightning fires that used to spread into the natural areas from outside are now usually extinguished, and yet another source of natural fires is lost. Even in the largest natural areas, manager-ignited prescribed fires may be necessary to restore fire to a semblance of its historical frequency (Arno, Parsons, and Keane 2000, Zimmerman and Bunnell 2000). Small wilderness and natural areas will almost certainly have to rely on some prescribed fires to maintain the natural fire process (Brown 1993).

Our Progress Report

Despite widespread agreement among forest management professionals that fire is a necessary ecological process in western forests, management still relies heavily on attempts to exclude fires without adequately substituting prescribed fire, silvicultural cuttings (as opposed to timber extraction), and other fuel treatments. Some progress has been made since the 1970s in restoring some fires to a few wilderness and national park areas, and prescribed burns are improving substantial acreages in general forest management zones on federal lands. We now have the technical knowledge to manage forests to reduce wildfire hazards and to restore many of the ecological benefits associated with the natural fire regime. We now need the political will to act on this knowledge. Progress is limited by reluctance to fund restoration even if it is more cost effective and productive than attempts to exclude fire. While some people are suspicious of the motivations behind any activity that generates revenue, the practical fact is that unless Congress significantly changes funding formulas for fire management, fuel reduction projects must be economically viable as well.

Although we recognize that the century-old fire suppression policy best described as attempted fire exclusion has failed to maintain healthy forests or protect private and public property, we are still a captive of this concept and the bureaucracy it has spawned. Federal and state spending on fire suppression makes it one of the major economic activities in the inland West (Devlin 1995). One environmental critic calls it the "fire industrial complex" (The Economist 2001, 33). For decades, fire suppression efforts have been explicitly modeled on military operations, with fire as the enemy. Fire suppression has a vital role in forest management, just as military preparedness has in national security. However, suppression should no longer dominate our management of fire and fuels in western forests. Instead, suppression should serve as a tool that keeps unwanted fires at bay but also allows us to restore some of the benefits of natural fire.

Some people argue that since forestry leaders brought us the fire exclusion policy a century ago, there is little reason to follow their advice now and adopt a radically different policy. However, the idea of adapting to a fire-prone environment is traditional in preindustrial societies worldwide (Pyne 1982, 1997). Fire exclusion should rightly be considered the radical, unproven, even "unnatural" approach. Scientific studies of the natural role of fire in maintaining forest communities began long after the fire exclusion policy was established. By the 1960s, ecology had emerged as a major branch of science, and its findings reinforce the concept that humans can help maintain natural systems by adapting to and working with them rather than imposing drastic alterations on them.

Chapter 10

Restoring Nature's Creative Force

Two years after the Canal Fire burned in the Wallowa-Whitman National Forest of northeastern Oregon, the landscape was an eerie study in contrasts. A previously rich, productive, mature forest dominated by Engelmann spruce and lodgepole pine had been completely killed by a high-intensity wildfire, probably not unlike fires in past centuries. The trees still stood, their branches sticking out as partially consumed stubs. Light splotches where woodpeckers in search of bark beetles had hammered off the bark punctuated the blackened trunks. But on the ground, a luxuriant growth of grasses and flowering herbs had risen literally from the ashes.

In the midst of this thoroughly charred forest landscape, a small square patch of surviving living forest stood incongruously. The conflagration had passed by this stand of young 30-foot trees. In fact, there was scarcely any sign of burning at all. No doubt firefighters could have taken refuge here while the spectacular crown fire roared through the surrounding forest. Why did these trees escape? The small island of green was a former clearcut where the logging slash had been piled and burned. Its trees were not crowded together, and there was only sparse fuel on the ground. This patch of fuel-deficient forest surrounded by a stand-replacement fire testifies to how remarkably effective mechanical treatment of fuels can be.

Collecting and burning logging slash is just one kind of fuel and fire treatment that could be used to accommodate fire in our western forests. While there are many ways to reduce fuel and apply fire beneficially, every treatment to return an approximation of nature's creative force has some risk, may be hard to do, or may have perceived drawbacks such as changing the landscape visually. While we continue to delay or reject fuel management, a beguiling alternative remains in effect. Keeping the status quo requires no thinking and no responsible action, costs nothing now, and causes no disturbance. If we value wild places free of human influence, we can take heart that we are "leaving nature alone." This alternative is commonly called "no action" in forest planning, but continuing the attempt to exclude fire from the forest without providing substitute treatments is in its own right a significant intervention in the forest ecosystem. The alternative to continuing to keep fire out, cloaked as "no action," is to work just as actively to restore natural fires. The ability to use natural fires increasingly depends on creating hazard reduction zones to reduce risks while allowing fire to run its course. How recent fires burned in previously logged, burned, and untreated forests points to several creative approaches to restoring fire.

The "No Action" Alternative

Several years ago a regional forester for the U.S. Forest Service remarked in obvious understatement that he would hate to have to prepare an environmental impact statement (EIS) justifying the fire suppression policy. He was acknowledging the elephant in the living room that has haunted American forestry for nearly a century: Although the policy of attempting to exclude fire was established to protect forests, suppression activities frequently cause damage instead. In addition, keeping fire out of fire-dependent forests may allow them to deteriorate and may increase the hazard of damaging fires.

When evaluating options for managing publicly owned forests today, many people choose the "no action" alternative in preference to any kind of cutting, fuel treatment, or prescribed burning, assuming that this choice will preserve the forest as they are accustomed to seeing it. "No action" also happens by default, as the result of not making a choice. Virtually any treatment on public lands requires a formal environmental analysis (EA) or EIS. The "no action" alternative continues to exclude fire without employing substitute treatments to mimic the effects of natural fires. Since the decline of logging in the 1990s, "no action" has become the most common practice on many federal forest lands. "No action" departs radically from natural processes and is likely to degrade the forest as a habitat for the historical native plants and animals (Arno 2000, Lyon and others 2000). Yet no formal document is necessary to implement this alternative on public lands.

The irony runs deeper. Restoration treatments place responsibility on the land manager. If something goes wrong, such as smoke from a pre-scribed fire descending on a community, or worse, an escaped fire burns private property, he or she may face bitter criticism or worse. In contrast, the land manager can choose the "no action" alternative, or activate it by procrastinating and not making a choice, with little concern about being held accountable. Even if "no action" leads to a disastrous insect outbreak, disease epidemic, or wildfire, most people will accept this as an uncon-trollable act of nature. Many dedicated conservationists argue for "no action" to save the forest because they have fought against industrial log-ging for so long that they have little reason to trust government land man-agement. The "no action" alternative also fits well with the classical steady-state picture of ecology that minimized the role of disturbance and change in natural ecosystems (Botkin 1990).

Juggling Natural Fires

In 1988 a national uproar was fueled by television news reports of destruction being wreaked by "let it burn" fires in Yellowstone National Park. President Reagan and many other political leaders spoke out against the seemingly illogical policy of allowing fires to burn in a cherished nat-ural area. Programs allowing natural fires to burn had been operating suc-cessfully for many years in Sequoia, Kings Canyon, and Yosemite National Parks in California; the Gila/Aldo Leopold Wilderness complex in New Mexico; and the Selway–Bitterroot Wilderness 200 miles west of Yellowstone. However, the national media seemed unaware of these nat-ural fire programs, which had attracted only limited, local attention. Sur-veys of backcountry users in the Selway–Bitterroot Wilderness showed a high and growing approval of allowing natural fire to return (McCool and Stankey 1986). Criticism of fire management in Yellowstone halted most natural fire programs in the West for two years. It also led to high levels of caution when the programs resumed, reducing the manager's ability to restore natural fires to their historical role even in large wilderness areas (Agee 2000, Parsons 2000).

The history of the natural fire program in the Selway–Bitterroot Wilderness illuminates the consequences of the post-1988 setback and suggests what might be done to regain the lost ground. The Selway–Bitterroot is the remote and rugged backcountry where pioneer western forester Elers Koch (1935) first lamented the failure of the Forest Service's policy of attempting to eliminate fire. Despite Koch's objections, agency administrators intensified fire exclusion. Finally in 1971, Bud Moore, Orville Daniels, and Bob Mutch, foresters who had spent much of their careers in the vicinity of the Selway–Bitterroot, were given the chance to allow natural fire to return to this wilderness ecosystem (Devlin 2000, Moore 1996).

Moore and his colleagues implemented a plan that allowed lightning fires to burn in a specific area of the Selway–Bitterroot under previously determined weather and fuel conditions (Devlin 2000). In the summer of 1972, when this natural fire program was instituted, lightning lit one small fire in the experimental area, but damp conditions put it out. The first real test came in August 1973, when two fires in the White Cap drainage were allowed to burn. They grew rapidly, and hand-built firelines had to be constructed to keep them from spreading outside the experimental area (Devlin 2000). Some foresters, firefighters, and area residents considered this operation sheer foolishness and roundly criticized Moore and his collaborators. After all, the U.S. Forest Service initiated and led the nation's aggressive program to exclude fire from forests. The act of allowing a fire to burn freely in the forest contradicted the agency's avowed mission and the training, philosophy, and experience of its staff (Daniels 1991). At that time only Sequoia and Kings Canyon National Parks in California had allowed any natural ignitions to burn (Kilgore and Briggs 1972). Nevertheless, by the time rains quelled the blazes in the Wilderness in late September, a total of 2,800 acres had burned, and the precedent for allowing natural fire to return to the landscape had been set.

During the next few years, the natural fire program was expanded to larger portions of the Selway–Bitterroot, and by 1979 the entire 1.3-million-acre Wilderness was included in the plan. Since then, fires caused by lightning within prescribed conditions have been managed for their benefits as natural fires. Most fires caused by humans have been managed under standard firefighting procedures. A "confinement" strategy has been used on many of the fires caused by lightning outside of prescribed conditions (Brown and others 1994). This employs suppression only as needed to keep fires from escaping the Wilderness or threatening facilities. These fires contribute to the overall goal of returning natural fire to the ecosystem. During busy fire seasons, new lightning ignitions in the Selway–Bitterroot have often been suppressed to keep fire activity within manageable levels. A similar natural fire program was soon implemented on the 2.1-million-acre Frank Church/River of No Return Wilderness immediately south of the Selway–Bitterroot in central Idaho (Benedict, Swan, and Belnap 1991).

Between 1979 and 1990, a total of 150,000 acres in the Selway–Bitterroot burned in natural fires and fires with limited suppression (Brown and others 1995). This is about 60 percent as much burning as occurred in the average 12-year period prior to the initiation of fire suppression, based on fire history studies (Brown and others 1994). However, the area burned under the natural fire program is a vast increase compared with the period of effective fire suppression from 1935 to 1972. In the ponderosa pine forest at the lowest elevations, fires occurred prior to 1935 at average intervals of about 22 years at any given location (Brown and others 1994). During the natural fire program (1979 through 1990), fires occurred less frequently in the ponderosa pine type, at a rate of burning that would produce average intervals of 59 years. In the high-elevation whitebark pine type, the fire intervals were also substantially longer under the natural fire program than in the historical period before fire suppression. All the other forest types had rates of burning under the natural fire program that were roughly comparable to the historical rates. The program of introducing fire, though not completely adequate, did show a measurable degree of success in restoring fire to historical levels (Benedict, Swan, and Belnap 1991, Brown and others 1994, Daniels 1991).

Poor air quality from forest fire smoke is a common reason given for limiting prescribed burning. However, there was no real difference in episodes of forest fire smoke in the populated Montana valleys downwind from the Selway–Bitterroot Wilderness during the full fire suppression policy (1960–1971) compared to the natural fire policy (1979–1990) (Brown and others 1995).

After the 1988 Yellowstone fires, the U.S. Departments of Agriculture and Interior conducted two formal reviews of the natural fire policy (Agee 2000). One panel reviewed the natural fire policies of Yellowstone National Park (Christensen and others 1989), and the other examined national fire policy (USDA-USDI 1989). Reviews found that the policy was generally sound, but in Yellowstone National Park and several other areas more specific criteria were needed for determining when fires should be allowed to burn. Most of the suspended natural fire programs resumed, but national park and wilderness managers set more conservative criteria for allowing natural fires to burn. Importantly, all these criteria targeted fire control or smoke concerns and none aimed to return the historical role of fire to the ecosystem (Agee 2000). The predictable result throughout the West has been fewer acres burned by natural fires since 1988 (Parsons 2000, Parsons and Landres 1998).

After the 1988 wildfire season the Forest Service aggressively suppressed most natural fires in the eastern one-third of the Selway–Bitterroot Wilderness. Land managers feared that a fire leaving the eastern boundary of the wilderness might threaten forest homesites in the adjacent Bitterroot Valley. Forest residents also complained to the agency when they saw smoke in the nearby wilderness area. With people building more and more homes in forests adjacent to wilderness areas since 1988 this political pressure continues to increase.

In spite of a decline across the West in natural fires since 1988, the National Park Service has successfully implemented a natural fire program in the northwestern portion of Glacier National Park. The broad, heavily forested valley of the North Fork of the Flathead River historically experienced a mixed fire regime, with fire intervals at a given location averaging 25 to 75 years (Barrett, Arno, and Key 1991). After the severe Red Bench Fire in 1988, park managers suppressed natural fires only to confine them to areas away from facilities. Between 1994 and 2000 these fires burned about 37,000 acres in mixed burning patterns re-creating burn mosaics similar to those of historical fires (Kurth 1996, Vanhorn 2001).

Hazard Reduction Zones

Many residents east of the Selway–Bitterroot Wilderness have actively opposed Forest Service proposals to reduce forest fuels outside the wilderness boundary using selective harvesting, thinning, or prescribed burning. Opponents don't want to lose their view of what they value as an untouched forest. They are also concerned that treatment will result in more roads or all-terrain vehicle use. At the same time, homes in this area remain vulnerable to fire as long as there is little fuel reduction.

During the 2000 fire season more than 500 families were evacuated west of Hamilton, Montana, when the human-caused Blodgett Fire threatened their homes. The ponderosa pine–fir forests surrounding these homes historically burned in frequent low-intensity fires. As long as fuels management treatments, prescribed burning, and natural fire are not politically acceptable, the only management option available (fire exclusion) will alter the natural diversity of the Selway–Bitterroot Wilderness and actually increase the hazard of wildfire to nearby homeowners. Fire officials, forest managers, county administrators, and political leaders fearing a replay of the 2000 fires have joined in urging forest homeowners to remove fuels around their houses and to support fuels management on nearby public lands. In spite of growing interest, many residents lack the basic knowledge or skills to carry out fuel treatments around their homes. There might be a lack of motivation too. People may think the government and its firefighters will be able to protect them. After all, the Forest Service went to extraordinary ends to save homes exposed to the Blodgett Fire and other 2000 blazes. On Blodgett, 12 heavy-lift helicopters were employed at about $2,000 per hour to dump water on homes surrounded by heavy fuels. Agency officials warn that such elaborate, publicly-funded fire protection is unlikely to be available in the future.

How might broad-scale fuel treatments be accomplished in western forests? About 39 million acres of national forest land in the interior West are at high risk of catastrophic wildfire as a result of 70 to 100 years of fire suppression (GAO 1999). Carrying out fuel reduction treatments on such a vast expanse of forest would be virtually impossible in any reasonable period of time. However, even if only the 2 million acres most strategically located near forest homes and developed areas were treated, the costs of fighting future fires and the amount of wildfire damage to private property might be reduced by 30 percent or more (Jones and Chew 1999).

FORESTED FUELBREAKS

Creating forested fuelbreaks is a logical first priority treatment to reduce the probability of severe wildfire. Fuelbreaks provide immediate wildfire protection and also serve as starting points for broader fuel treatments. A buffer zone about one-quarter-mile wide consisting of open forest of moderately large trees with only light fuels can greatly subdue wildfire intensity (Agee and others 2000, Anderson and Brown 1988, Scott 1998). Wildfires will burn through these forested fuelbreaks, but they are typically reduced to surface fires of low to moderate intensity that many of the fire-resistant trees like ponderosa pine or other species with thick bark will survive. No fuel-treated zone is absolutely effective in all situations, but a buffer zone can give firefighters a chance to prevent damage to property in the lee of the fuelbreak, including homes and developments.

Forested fuelbreaks lower fire intensity by reducing the amount of surface fuels and ladder fuels, and by increasing the distance to the bottom of the forest canopy and the amount of open space separating tree crowns (Agee and others 2000). To create a fuelbreak in most western conifer forests, it is necessary to thin the trees, removing most of the smaller trees, which represent ladder fuels or have branches extending close to the ground. Additionally, enough of the taller trees should be removed to reduce crown cover to less than 35 percent with a minimum of 10 feet of open space between tree crowns (Schmidt and Wakimoto 1988). Crown cover is the percentage of total area that is covered by tree canopies when viewed from above. This spacing is more open than many historical stands or what we propose for broader fuel treatments in the forest, but it applies only to a relatively small area. Where possible, it is best to leave mostly the fire-resistant trees.

The stems of larger trees can be removed as commercial timber for sawmills or as pulpwood or firewood. Treetops, branches, and saplings might be chipped and sold as fuel for an industrial boiler or electric-power generator. If chipped material is left in the woods, it should be scattered and not left in piles due to the danger of spontaneous combustion. Small stems, tops, and branches might be burned in piles. Accumulations of dead brush, old slash, and dead trees should also be removed or burned. If there are remaining trees with low-hanging canopies, the branches should be pruned off up to a height of at least 10 feet. Costs, though substantial, may be moderate in relation to costs of fighting wildfires in unmanaged fuels and costs of wildfire damage, each of which commonly run over $1,000 per acre.

Forested fuelbreaks should be accessible by road to allow maintenance of the vegetation and to make them available for wildfire suppression. Roads also serve as a narrow fuel-free fireline within a forested fuelbreak. Low-impact, off-road forest harvesting equipment is becoming widely available and can be used in constructing fuelbreaks that are not road-accessible. Certain topographical positions such as stream courses or ridgetops are most favorable for fuelbreaks because fire spread may already be slowed somewhat by increased fuel moisture or unfavorable conditions for preheating of fuels (Agee and others 2000). Maintenance every few years will be necessary to reduce thickets of saplings and concentrations of dead tall shrubs. In some situations prescribed burning in late winter, when the adjacent dense forest is moist, may be possible.

AREA-WIDE FUEL REDUCTION

Forested fuelbreaks do have fewer trees per acre than the rest of the forest can sustain. On a larger scale, the most useful approach to maintaining low-hazard forest conditions is "uneven-aged management." This is designed to perpetuate an open-grown forest with trees of many ages and sizes, including old growth trees if desired (Arno and Harrington 1998). Area-wide treatments can be designed in patchy patterns on the landscape to maintain areas of heavy cover for wildlife habitat and biological diversity (Camp, Hessburg, and Everett 1996, Weatherspoon and Skinner 1996). In fact, the fueltreated areas can help protect areas of dense forest cover from severe wildfires (Jones and Chew 1999). Over long periods of time, new areas providing dense cover can be allowed to develop, while old ones are treated and recycled. This simulates the shifting mosaic of stand ages and structures across the entire landscape. While no single patch of ground will have the same stand composition on it forever, the entire mosaic will retain its basic character.

Uneven-aged forest management usually requires treating the stand at intervals of 20 to 35 years to maintain open conditions and a balance between the numbers of young, intermediate, and older trees (Fiedler, Becker, and Haglund 1988, Guldin 1995). Small, medium, and large trees in excess of those needed for the future stand are removed each time. Most treetops and other slash should be removed, burned in piles, or burned in place in a controlled surface fire. Uneven-aged management focused on fuel reduction and ecological restoration can return a semblance of historical conditions in forests that were characterized by understory fire regimes or mixed fire regimes, and therefore had stands with fire-resistant trees of different ages (Arno, Smith, and Krebs 1997, Hartwell, Alaback, and Arno 2000). This management also benefits the herb and shrub species that were historically associated with these fire regimes.

Historically, frequent fires in ponderosa pine and pine–fir forests often perpetuated stands with 30 to 100 trees per acre, mostly of medium and large sizes. Today's forest at the same spot is likely to have accumulated 50 to 100 percent more overstory trees along with hundreds or even a few thousand small trees per acre. This modern forest structure represents a buildup of ladder fuels as well as continuous canopy fuels. Because the volume and biomass of living trees is often double that of historic forests, competition for the limited supply of soil nutrients and moisture is intense. This results in extremely slow growth of even the larger trees and stagnation of tree growth throughout the forest (Arno, Scott, and Hartwell 1995, Fiedler 2000a). The small, understory trees compete aggressively for the scarce moisture and nutrients (Biondi 1996). If left alone, most of today's dense wildland forests will probably never be able to produce large trees like their historic counterparts, due to excessive competition. Similarly, many of these forests will become dominated by shade-tolerant trees like western hemlock, white fir, or inland Douglas-fir at the expense of the historically dominant, long-lived, fire-dependent trees, such as ponderosa pine, Jeffrey pine, sugar pine, western white pine, western larch, and coastal Douglas-fir (Hartwell, Alaback, and Arno 2000, O'Laughlin and others 1993). The lack of tree vigor increases the severity of epidemics of bark beetles, defoliating insects, dwarf mistletoe, and root rots (Anderson, Carlson, and Wakimoto 1987, Filip 1994, Monnig and Byler 1992). The dense tree canopy stifles growth of many herbs and shrubs that are important for wildlife habitat and contribute to biological diversity.

Unlike forest management designed to achieve the greatest economic return from timber production, uneven-aged management for ecological restoration has primary goals of maintaining low wildfire hazard and simulating historical forest conditions of biological diversity (including wildlife habitat) that result from natural ecological processes. The forests that result from this management provide wildlife for viewing and hunting, other recreation, and timber products (Arno and Harrington 1998, Becker and Corse 1997, Covington 1996, Fiedler and Cully 1995, Guldin 1995, Smith 2000c). Unlike industrial plantation-style management, this uneven-aged management aims to retain a more natural forest community over the long term while still providing multiple-use benefits. The emphasis during every treatment is on what trees should be left, not what can be extracted. Nevertheless, when the appropriate treatments are made to achieve the above goals, the economic value of timber removed is often sufficient to pay for the entire treatment (Fiedler and others 1999, 2001).

Only a small percentage of the forest landscape will receive a restoration treatment within the next 20 years because of the scale of the problem (Weatherspoon and Skinner 1996). Selection of areas for treatment could be prioritized with the following criteria in mind. The area is (1) accessible from existing roads, (2) located near forest homes or other developments, (3) strategically situated in the landscape so that treatment would benefit a broader area, (4) at risk of a relatively high wildfire hazard, (5) in need of restoration for ecological reasons, and (6) conducive to restoration treatments being economically feasible and allowable in terms of the area's management goals.

Figure 10.1. A managed, uneven-aged ponderosa pine–fir forest. Note the mixture of different-sized trees, generally open understory, and occasional patches of young pines. This is more similar to historical conditions than many unmanaged forests display. (USDA Forest Service photo.)

In ponderosa pine–fir forests, restoration often involves heavy thinning of understory trees and uneven-aged management of the overstory to reduce density to historical or sustainable levels. Overstory cutting concentrates on removing weaker trees and firs and leaving a balanced quantity of the more vigorous trees in all size classes (Fiedler 2000b). Figure 10.1 shows the results of this type of treatment. After the initial cutting and fuel reduction treatments, including burning, have been done, maintenance treatments are made every 20 to 30 years or so, and these could rely more heavily on prescribed burning.

Initial restoration treatments using uneven-aged management in ponderosa pine forests have been tested in the Bitterroot National Forest (Smith 2000c) and are being conducted or proposed in the Colorado Front Range (Kent, Shepperd, and Shields 2000), Oregon's Blue Mountains (Starr, McIver, and Quigley 2000), California pine forests (Kiester 1999, Oliver 2001, Zack and others 1999), and northern Arizona (Moore, Covington, and Fulé 1999) including the Grand Canyon Forests Partnership (Gerristsma 2000). Silvicultural cutting and fire for initial restoration in natural areas have been tried in ponderosa pine forests of Grand Canyon National Park (Heinlein and others 2000) and pinyon–juniper woodlands in Bandelier National Monument, New Mexico (Sydoriak, Allen, and Jacobs 2000). Experiments using prescribed burning alone for restoration have run over many years in giant sequoia groves in Sequoia and Kings Canyon National Parks (Keifer, Stephenson, and Manley 2000, Stephenson 1999). Other approaches for ecological restoration are used to maintain dense-cover habitats in the Cascade Range (Camp, Hessburg, and Everett 1996), fire-dependent streamside forests (M. K. Arno 1996), Oregon oak woodlands (Agee 1996, Tveten and Fonda 1999), coastal Douglas-fir forests (Means, Cissel, and Swanson 1996, Means and others 1996), lodgepole pine forests (Zimmerman and Omi 1998), and aspen woodlands (DeByle and Winokur 1985, Duchesne and Hawkes 2000). There is considerable acceptance of these treatments from the community groups that have participated, but some critics argue that any tree cutting or any commercial use of trees is unnatural or will ultimately lead to a resurgence of industrial logging on public forests.

Forest areas intended for dense cover and places such as high-elevation moist forests that historically had a stand-replacement fire regime might be fuel treated using "even-aged" management, where most of the trees are cut or burned to start a new stand. This can be done by cutting some of the trees in irregularly shaped areas and strips in conjunction with prescribed burning (Arno 2000, Franklin and others 1999, Hardy, Keane, and Stewart 2000, Means, Cissel, and Swanson 1996).

Considerable knowledge is available to conduct the prescribed burning component of restoration treatments (Biswell 1989, Hardy and Arno 1996, Kilgore and Curtis 1987, Martin 1990, Sackett, Haase, and Harrington 1996). During the 1980s and 1990s, tens of thousands of acres were burned per year in conjunction with uneven-aged management in western national forests (Bailey and Losensky 1996, Kilgore and Curtis 1987, Simmerman and Fischer 1990). Many different approaches are available, including manipulation of fuels or varying the pattern of ignition, time of day, weather conditions, and fuel moisture. Still, it is clear that such practices need to be expanded greatly to fully meet fuel reduction and forest restoration needs across the West (GAO 1999).

In some of the areas that have dense forests as a result of fire suppression, it is not economically feasible, socially acceptable, or permitted by regulations (as in wilderness) to use mechanical thinning or other cutting treatments. In those situations prescribed burning appears to be the only tool available for restoring more natural conditions. However, profound changes in forest conditions often make it difficult to restore fire without first removing the buildup of trees or fuels (Harrington 2000, Heinlein and others 2000, Miller and Urban 2000, Pyne 2001b). Surface fuels have often accumulated and most trees are growing very slowly in a stressed condition. Low-intensity fire will not kill enough trees to adequately thin the stand because many are too large and fire-resistant. Conversely, if the fire is sufficiently intense to kill many trees, it is likely to kill or weaken most of the stand. Successive, moderate prescribed fires might eventually restore a more desirable forest structure.

But using fire alone in ponderosa pine and other forests where several natural fire cycles have been missed is inherently risky (Pyne 2001b). The disastrous result of the prescribed burn that escaped Bandelier National Monument in May 2000 and eventually burned homes in Los Alamos, New Mexico, demonstrates how dangerous it is to rely on fire alone to restore natural forest conditions. The most immediate factor responsible for the fire's escape was that a careful prescription and burn approval process was not followed (American Forests 2000). But even with better planning, with heavy fuel loads to contend with, managers have far less leeway in weather conditions that allow safe, controllable burning. As already described, it is much less risky to return fire to forests if excess trees and excess fuels, accumulated due to missed natural fire cycles, are removed first. Establishing a forested fuelbreak near a natural area boundary or around adjacent communities should be the logical first step in using prescribed fire.

Insights from Fire Effects

Since the early 1980s severe wildfires throughout much of the West have left huge 500-year-old ponderosa pine trees standing grotesquely with ashy gray foliage still "frozen" in the direction of the inferno's superheated wind. Most of these fire-killed trees survived anywhere from 12 to 80 understory fires in their first four centuries of existence, but then there were no fires in the last 80 years or so. Although historical fires killed a small proportion of these ancient trees, modern wildfires often kill most of them. They are victims of our current fire exclusion policies. Wildfires are killing the old growth pines in stands that were selectively logged as well as in natural areas where no logging occurred.

We can see how to sustain forests by looking at the effects of recent wildfires in stands that had different kinds of timber harvesting and fuel treatments. Two types of harvesting were most common in the ponderosa pine–fir forests that burned in the 2000 fire season. First, clearcutting in the 1960s and 1970s gave rise to dense, even-aged stands of planted ponderosa pine mixed with some other conifers. By 2000, these trees were 15 to 25 feet tall and had live or dead limbs near the ground, with an abundance of dry grass and shrubs beneath. Second, selective cutting of large trees in past decades allowed thickets of small trees, limbed to the ground, to develop beneath the remaining tall trees. Both of these stand structures provided ample ladder fuels and continuous canopy fuels. Both forest structures contrast markedly with the most common historical condition in ponderosa pine–fir forests, which was open-growing stands of trees with few low branches and sparse understory fuels. By the mid twentieth century Harold Weaver and Harold Biswell were recommending uneven-aged management coupled with prescribed burning to keep trees well spaced and prevent a buildup of understory fuels. Unfortunately, until recently the prevailing viewpoint in western forestry was that clearcutting or selective cutting without prescribed burning were more efficient, and that our highly capable and well-equipped firefighting crews could control most wildfires regardless of conditions.

Even some recently thinned ponderosa pine–fir stands with open understories were killed in the 2000 wildfires in the Bitterroot National Forest and state forestlands. Their fate shows that thinning without periodic fuel reduction is not adequate to protect a stand from fire. In some of these stands a moderately heavy load of slash had been left to enhance cycling of nutrients into the soil. In other thinned stands, forest floor litter and duff and other fuels built up over many decades may have allowed high-severity surface burning. Although the stand structure was closer to the historical norm, there was still much more fuel than the clear ground that the early pioneers first encountered. Indeed, how could fuels accumulate when fires used to occur at average intervals of 5 to 30 years, or even less in some parts of Arizona and New Mexico? One rancher near Sula in the upper Bitterroot Valley had previously thinned his forest and burned the slash in piles. In August 2000, stand-replacing fire roared into his forest from the adjacent state land, but soon decreased to an underburn that most trees survived.

Treatment also may not succeed if it is applied to too small an area. During the big blowup of the 2000 Valley Complex fires in the East Fork Bitterroot drainage, high-intensity burning engulfed some stands that were not dense or heavy-laden with dead fuels. This mortality can be expected especially in small stands upslope from or surrounded by dense forests with built-up fuels. Similarly, fuel-treated stands within central Washington's Tyee Fire (1996) experienced heavier than expected mortality due to the vast expanse of surrounding untreated forest that burned in crown fire (Agee 2001). The crown tops of the treated stand were scorched from hot air from nearby crown fires. Then the crown fire shifted to an underburn in the treated stand, scorching the lower crown. Even though the middle of the crowns remained green, many of the trees were weakened enough to allow bark beetles to kill them.

Previous fires and logging also influenced fire behavior at middle and upper elevations during the 2000 fires, especially in the lodgepole pine–subalpine fir forests where stand-replacement burning was common historically. The 28,000-acre Sleeping Child Burn of 1961 lay directly in the path of intense burning on the Bitterroot National Forest. By 2000 it was well stocked with young lodgepole pine 20 to 30 feet tall. Much of the 1961 burn had been salvage logged, and the 2000 fires generally burned lightly or not at all through it. In contrast, the old forest that surrounds the 1961 burn experienced a great deal of stand-replacement burning. This reinforces the concept that fuel treatments can reduce the hazard of stand-replacement burning in lodgepole pine forests (Omi 1996, Omi and Kalabokidis 1991). This could be useful for protecting resorts, summer homes, and other high mountain facilities. Even in unin-habited areas, it may be desirable to break up continuous heavily fueled lodgepole pine forests to re-create the historical mosaic and to reduce the scale of wildfire impacts on high mountain streams or forest land provid-ing critical habitat for certain fish or wildlife species.

Creative Concepts for Forest Restoration

On most federal lands in the West, we continue trying to keep fire out of fire-dependent forests while not carrying out appropriate substitute treat-ments. Some timber harvesting operations have included fuels treat-ments, such as removal of excess understory trees and prescribed burning. However, timber harvesting usually does not also treat fuels under a long-term perspective. If, in the early part of the twentieth century, we had fol-lowed the advice of the timbermen and local residents who advocated light burning in the forest, we might now have figured out how to main-tain fire-dependent forests on much of the landscape. Today, a buildup of forest fuels and high-value forest homes and developments at risk makes forest restoration harder. Historically dominant tree species such as pon-derosa pine, sugar pine, western white pine, and larch have often been dis-placed as a result of selective logging and suppression of natural fires (Hartwell, Alaback, and Arno 2000, O'Laughlin and others 1993).

In spite of these obstacles, we do have much improved technical skills and equipment. We can remove, and even utilize, small trees more eco-nomically than in the past, and our fire control technology allows application of prescribed fire in innovative ways. A restoration thinning treatment can create slash that will carry fire in late winter or early spring when the duff and other natural fuels will scarcely burn. After the slash is burned, we can burn again to reduce the natural fuels. We can use aerial ignition to burn efficiently over large areas or in rugged, inaccessible ter-rain (Intermountain Fire Council 1985, Martin 1990). We can burn after autumn frost creates additional fine fuel, or in late winter on dry south-facing slopes that are surrounded by snowy or moist forests (Keane, Ryan, and Running 2000).

Still, the amount of prescribed fire we can realistically apply is limited by the small number of days per year that have both favorable weather and fuel conditions, availability of fire personnel and equipment, and concerns about risk and smoke pollution. We need to use prescribed fire where it can do the most good and can be applied efficiently. Often this will require pretreatment of excessive fuels by mechanical means. In many situations near residential forests, fuel reduction is desperately needed and will have to be done by mechanical means alone. Low-impact, all-terrain harvesters manufactured in Finland can collect and compress small understory trees, limbs included, in compact 1,200 pound "energy bundles" for use in generating electricity and steam to heat schools or other large buildings (Uuskoski 2001). Each bundle produces about 1.2 megawatts of energy. Discovering the best methods of harvesting and utilizing excess forest fuels is critical for achieving fuels management on a significant scale.

Computer modeling, illustrated in Figure 10.2, can help determine

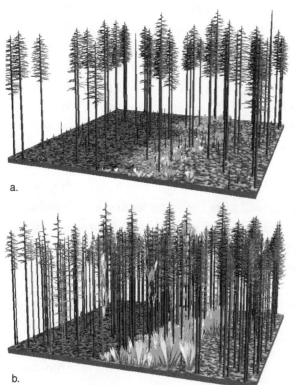

a.

b.

Figure 10.2. Computer simulation of fire behavior and tree mortality in a second-growth ponderosa pine and Douglas-fir stand. The top image (a) shows the stand 10 years after the second of two treatments of thinning and prescribed burning spaced 20 years apart. Even under extreme fire conditions of high heat, wind, and low humidity, this stand only sustains a low-intensity underburn. The bottom image (b) shows the same stand after the same time period but without any treatment. It burns in a crown fire under extreme fire conditions. (Fire FX module of the Forest Vegetation Simulator developed by the USDA Forest Service.)

patterns and kinds of fuel treatments that would most effectively reduce wildfire hazard to homes and private property (Finney 2001, Jones and Chew 1999). When such treatment zones are in place, more remote lands behind them could be treated more liberally with prescribed fire or natural fires because of reduced hazard to developed areas. If we will finally notice the elephant in our living room, the ecologically destructive and economically costly effects of fire exclusion, we could justify shifting funding priorities toward fuel treatments and fire restoration.

Chapter 11

Managing Wildland Fuels around Homes

Would you ignore obvious, avoidable hazards and expose your home to forest fire? If you built a home in the middle of heavy forest fuels, you may have done just that. When you chose that attractive, rustic-looking cedar shake roof, did you think you were roofing your house with kindling? If you did either of these things, you were not alone—hundreds of thousands of vacation cabins and residences were built in forests across the West in the last 30 years. Through the mid 1980s, even insurance companies seemed oblivious to the threat of locating highly flammable homes in heavily fueled forests. Only in the California brushlands (chaparral) did home losses from wildfires become so extensive that insurers took notice of hazards, set minimum fire-safe building standards, and refused to insure unsafe homesites.

By the late 1980s, home losses to wildland fire had escalated in western forests. Hundreds of residences burned during severe wildfire seasons, which come every two to three years. During the 1990s some insurance companies concluded that forest fire damage was a correctable problem for home safety, and they joined the effort to urge homeowners to reduce hazards. Some insurers refused coverage to those choosing to live under high-hazard conditions. If you live in the forest, you're probably just as anxious not to see the trees around you burn as you are to protect your home.

157

The Surviving Home in a Seared Forest

In August 2000, the East Fork of the Bitterroot River suddenly changed from a familiar forest to a strange environment. Instead of the usual bright midday sun and summer heat, a thick pall of smoke from fires burning in the surrounding mountains filled the valley. Two weeks earlier, a conflagration had swept through this part of the rugged East Fork drainage in a variable burn pattern. Green pastures, much of the stream-side vegetation, and some open fields with little fuel had scarcely burned. Patches of open-grown conifers had mostly underburned, and the trees were still green except for scorched lower branches. Dense patches of trees, especially on the steeper slopes, had burned at high intensity and were dead. Most homes survived, probably because of good road access for firefighters, the broken, discontinuous pattern of landscape fuels, and nearby river water available for emergency sprinkler systems.

Up the narrow canyon of a steep side drainage, the burn pattern changed with a different kind of environment. The dense forest of 100-foot-tall conifers was virtually all black and stripped of foliage. The ground was black, and nearly all the duff and humus had burned. Here, all that remained of homesites were charred foundations, stone chimneys, and the occasional burned vehicle, although most of the cars, like the people, had been evacuated. Only four homes survived the inferno, but each was surrounded by black, incinerated forest. Three of these survivors had small openings, perhaps half an acre, with a watered grass field next to the house, and they had metal or other noncombustible roofs. Each of these homes had a pond attached to the creek that fed a sprinkler system.

The last surviving home stood near the head of the drainage, a small vacation dwelling with a metal roof. A lack of vegetation or combustible material within 30 feet of this residence spared it from the crown fire raging overhead. The property owner had bought this tract of forest only a few years earlier and had hoped to vacation and perhaps to retire here. But the devastating scene was a far cry from the green forest that he had purchased to enjoy. He had been aware of forest fire danger and had hoped to thin the trees. Now he weighed removing much of the dead timber and planting seedlings. He acknowledged his good fortune that the residence had survived, but he was discouraged that his woodlot, once green and beautiful, was now a dismal mass of burned trees and charred ground. He was faced with the unhappy task of removing the dead trees all around his dwelling before they might topple, and taking out some of the more distant trees lest the accumulation of dead wood fuel another fire in the future. He knew the forest would eventually recover. Still, this process, especially restoring good-sized trees, would take so long that he questioned whether he wanted to keep the place. On the other hand, who would buy it in this condition? Another resident, surveying a similar scene, lamented that he would rather have lost the home, which could be replaced in a short time, than the forest, which could not. After the crushing losses in scenic and property values of the 2000 fire season, more people recognize that one of the best ways to safeguard your house from forest fire is to make the surrounding forest fire-resistant as well. Here are some methods for reducing wildfire hazard to both your forest home and your home forest.

At Home in the Fire-Dependent Woods

The first step in adapting to life in the fire-dependent woods is to reduce the vulnerability of a homesite to wildfire. Suburban homeowners know all too well that they must constantly prune and trim the grass, shrubs, and trees in their yards to prevent them from becoming rank and overgrown. When hundreds of thousands of suburbanites began moving into western forests in the 1970s, most believed that, as a natural place, the vegetation surrounding their new homes would look after itself. But owing to fire exclusion, the forest around them was not as "natural" as they thought, and without tending and management, it accumulated fuel that could lead to destruction.

Most kinds of wildland forest that are now dotted with homes once burned fairly often in understory or mixed fire regimes. These forests have trees with some degree of fire resistance that were able to survive low- or even moderate-intensity fires. Suppression kept fire out of most forests for many decades, and in many cases the larger fire-resistant trees were removed by logging. Now, thickets of small trees and surface fuels have built up and wildfires are larger and more intense (GAO 1999). If a homeowner wants to allow nature to take its course, this is likely to be a stand-replacement wildfire, which will destroy any vulnerable buildings as well. Firefighters cannot safely or effectively combat stand-replacement fires to save homes. If you want firefighters to be able to successfully defend your property, you must reduce hazardous fuels before the fire arrives. It also wouldn't hurt to have your own fire-suppression capability in case firefighters are unavailable, as when they are trying to protect many homes during a large fire.

Protecting the Homesite

Even if the surrounding forest is made fire resistant, low-intensity fires will be able to spread through it, so it is essential to surround the house and other buildings with an area cleared of any significant fuels. Live herbaceous vegetation and some deciduous native trees and shrubs do not represent appreciable fuel if kept free of dead material and old leaf litter. In contrast, shrubby conifers like the sprawling junipers commonly used in suburban landscaping are highly combustible. Scattered, isolated conifers are not a fire threat if you keep them trimmed of low branches, dead branches, and large accumulations of lichens, which look somewhat like moss and are highly combustible. You should clean up duff and leaf litter beneath yard trees every year. If you are concerned about maintaining soil nutrition, culture relatively nonflammable compost material, such as humus from decomposed leaves of deciduous trees, in a place well separated from the house, or use commercially available soil amendments.

The needles of ponderosa, Jeffrey, and other long-needled pines are highly flammable and require special attention. Rake them up around the house and other buildings every spring, and frequently clean them out of gutters and off the roof and the deck. Avoid highly combustible landscaping material like cedar bark.

To separate them from forest fuels, buildings should have no combustibles connected to them, such as piles of wood or lumber. If the house has wood or other flammable siding, the area beneath the exterior walls should be free of fuel, since this is a likely landing spot for firebrands. Also, when small firebrands land on the house, they must not come into contact with pine needles or other fine fuels on the roof, gutters, and deck.

You should promptly replace wood shingle roofs with noncombustible roofing such as mineral-surfaced asphalt or fiberglass shingles, tile, or metal. After the 1988 fires swept through the Old Faithful area in Yellowstone National Park, wood shingle roofs on several cottages were pockmarked with burn holes from showers of firebrands that ignited but were promptly extinguished by fire hoses. Some cabins were not drenched in time and burned from the roof down. Fires have spread from house to house through large suburban areas in California as embers landed on wood shingle roofs, despite the inability of the flames to burn through surrounding lawns and landscaping. Decks also warrant special attention because they are wooden structures attached to the house where pine needles and other combustible litter can accumulate. The area under a deck also must be kept free of fuel.

Good access and turnaround space for large fire trucks is important. During big wildfires, firefighters generally have orders to pass by houses that have narrow access roads threading through dense, dangerous forests. Access routes to your home should have forest fuels thinned and removed from both sides of the road, and they should be marked clearly with road signs.

Finally, you should have a water supply and fire tools of your own. You should not depend on the electric power grid to continue operating during a wildfire. A pond or swimming pool can be coupled with a gasoline or generator-powered pump and a farm-type irrigation system to thoroughly wet the house and surrounding area in advance of a wildfire. A pond or small water reservoir can also allow a fire truck to refill its tank. The household water system with outside faucets and plenty of garden hose kept in good condition is useful for wetting the homesite in advance of the fire. Effective spray nozzles and portable impulse sprinklers for watering a generous area around the house are essential. Tools should include garden and leaf rakes and a mattock, pulaski, or heavy-duty fire rake for digging fireline.

Protecting the Home Forest

Thirty or more years ago, western woodlots often sold at modest prices based largely on the value of their timber. In recent years, however, the opportunity to enjoy the beauty of a place in the forest has multiplied the value of accessible forest tracts far beyond any timber values. Not surprisingly, a forest tract burned by severe wildfire immediately loses a great deal of its market value and salability.

Restoration Methods

Given the reasons you purchased your forest property in the first place, you will want to protect your forest setting from wildfire and from epidemics of insects and disease. Managing the forest to keep it safe and healthy will require you to trade in your suburban gardening expertise for the skills and tools of an amateur forester. The natural fire regimes that shaped these forests have been irrevocably altered; nevertheless, it is likely you can restore a semblance of the natural forest conditions because many components of the forest ecosystem still do exist. You can do this by removing certain trees, reducing excessive fuels, and perhaps using prescribed burning. Forest restoration means tending the natural forest using an ecosystem-based management approach similar to what is embraced for managing national forests and some large private forest lands (Salwasser and Pfister 1994). As a forest homeowner, you can carry out restoration treatments without being heavily encumbered by bureaucratic regulations, as are managers of public lands, or by the need for profit, as are large timber companies. Especially if you live on your woodlot, you have substantial freedom to restore and maintain the forest.

Even if you do not know where to start, you can acquire the knowledge through forest stewardship training programs available at minimal cost in each state. Participants in this training prepare a stewardship management plan for their property. Stewardship programs are cosponsored by the state Agricultural Extension Service located at state universities (see the Appendix). The state Extension Forestry office will have information about stewardship training and other educational programs and materials available to you.

Also, each state's department of natural resources or state forestry office (Appendix) employs service foresters who are available to visit your property and give you some ideas about forest management. Service foresters can provide information about programs to help you carry out forest improvements such as thinning of nonmerchantable trees, tree planting, creating ponds for wildlife, and stabilizing stream banks. Each state has a branch of the national Tree Farm Program, a nonprofit, nongovernmental organization established in 1941 to help private landowners develop good forest management by providing educational opportunities and sharing information (Appendix).

Equipped with knowledge acquired from the forest stewardship program and other sources, you can inventory forest conditions, decide management goals, and choose practices to meet your goals. Even as a small landowner, your projects still must follow the Clean Water Act when you work around streams, and you must abide by local regulations on burning. Private forestry consultants can also give advice or draw up plans. Contractors are available to conduct forestry work such as commercial logging, thinning of nonmerchantable trees, and tree planting. A consultant can develop a contract and provide guidance to ensure that any contracted work is performed properly. It is wise to inspect previous work done by prospective consultants and contractors to ensure that they have performed the kind and quality of work that you want. It is also useful to get an evaluation of past work from the landowners involved.

Often you and your family can do your own work and thereby learn firsthand about your own forest and gain the satisfaction of improving it. If you enjoy gardening, horticulture, or small-scale farming, you will likely find it rewarding to tend your forest. If you hire a forestry contractor, as a minimum it is a good idea to inspect the contractor's work the first day and often thereafter. If timber harvesting will occur, you or your consultant usually will want to mark the "leave trees," or commercial-sized trees that are not to be cut. It is important that leave trees not be scarred and that logging equipment does not rut the soil.

Both forest restoration treatments and logging designed to maximize income will remove some trees and leave others. However, the results are dramatically different. In forest restoration, the most vigorous trees of all sizes, including numerous large trees, are usually left, while weak or excessive numbers of trees are removed and thickets are thinned with minimal impact on remaining trees or soil (Manning 2001). Restoration forestry gives special consideration to providing snags for wildlife habitat and appropriate amounts of coarse woody material on the ground for soil nutrition and other ecological purposes. In contrast, logging to maximize income focuses on rapid removal of the valuable trees with little concern for the stand or the conditions that are left behind (Jay, Arno, and Arno 1998). Depending on the stand structure and size of the area to be treated, a restoration treatment may return some income to the landowner, but it will not maximize income. On the other hand, restoration treatments will improve the condition of the forest in several ways and allow more vigorous and valuable trees to develop. Some of these trees can be harvested in future treatments at intervals of perhaps 20 to 30 years, while continuing to improve the forest.

UNEVEN-AGED MANAGEMENT AROUND HOMES

If the forest surrounding a home is to have low wildfire hazard and low risk of insect and disease epidemics, uneven-aged management is often the most useful approach. Uneven-aged management in fire-dependent forests attempts to balance the numbers of large (old), medium, and small (young) trees as well as to create relatively open growing conditions so that most trees of all ages are vigorous and so that stand density will not sustain a crown fire (Arno and others 1995, Fiedler 2000a, 2000b, Fiedler and Cully 1995, Guldin 1995). This management approach is well suited to the acreage surrounding homes because it will develop and perpetuate scenic and safe forests that contain large trees. Uneven-aged management also perpetuates stand structures that were often found historically in ponderosa pine, Jeffrey pine, and some other western forests containing long-lived, fire-resistant trees.

You can seek guidance regarding how many trees to leave in each size class locally through forest stewardship training (Appendix) and consultants who have experience in forest restoration, not just in commercial logging. At the beginning of your treatment program, it is unlikely your forest will have an adequate number of trees in all size classes. Consequently, the goal of the first uneven-aged management treatment is to encourage a more balanced size distribution in the future. For instance, if the stand has only a few scattered large trees, cutting may leave them and remove the excess of medium and small trees. The treatment should concentrate on leaving the most vigorous trees in each size class and in favoring the long-lived, fire-resistant species. Some patches may have no healthy trees and all may be infected with root rot or heavy dwarf mistletoe. It may be desirable to remove most or all trees in such patches, perhaps burn the openings, and then plant the preferred disease-resistant species. Reaching a well-balanced uneven-aged stand structure may take successive treatments over 20 to 30 years or longer.

Some patches of dense forest cover can be maintained for wildlife habitat or for visual screening, such as to hide the view of a road or neighboring house. However, these patches should not be near buildings. Dense cover for screening or wildlife can be made less hazardous by keeping it in large shrubs or in small bushy trees no more than 15 feet tall, and by removing most of the tall trees associated with patches of dense cover. This will reduce the opportunity for crown fire to develop and spread into the surrounding forest.

There are many ways to dispose of the trees you need to remove. Medium and large trees, commonly those 9 inches or more in diameter 4 feet above ground, can usually be sold at a profit to a lumber or plywood mill. Medium and large trees, standing and fallen, are also the most valuable as habitat for many species of animals, but this value should be balanced with the need to keep fuel hazards moderate. Many tons per acre of small, nonmerchantable wood are likely to be harvested in the first restoration treatment. Small trees sold to a pulp mill or for fuel used for generating electricity may yield enough income to partly offset costs of removing them. You can sometimes sell small trees to fencepost and pole plants or use or sell them for firewood. Options for selling small trees and slash vary widely from one locality to another and fluctuate over time. Inquire with state service foresters and local tree farmers who actively market small trees to find out about new markets.

FUEL TREATMENTS

Treetops, limbs, and small trees that are left after harvest, along with any appreciable quantities of old down trees, should be treated in some way to lower wildfire hazard. Common methods include harvesting and skidding whole trees to a roadside landing where branches and tops can be separated and burned in very large piles. This usually leaves an area of heavily scorched earth. Other methods of slash disposal include chipping and hauling it away and piling and burning. Slash piles made with heavy equipment often contain a lot of soil and thus burn incompletely, leaving unsightly remains and smoldering embers that can start wildfires when fanned by high winds weeks later. Handmade piles can usually be consumed cleanly with minimal impact. Some snags and medium- or large-diameter pieces of wood can be left in the forest for wildlife habitat or nutrient cycling without affecting wildfire hazard.

Burning small slash piles is a routine, low-risk operation if done well outside of the wildfire season and soon after snowfall or wetting rains, to prevent the fires from spreading if a strong wind arises. An inexpensive propane torch is a prompt and efficient means of getting piles to burn well despite cool, damp conditions. Responsible burning requires attention to burning permits, fire tools, water supply, and follow-up surveillance. Usually these are not difficult to arrange at your homesite. Burning within a few hundred feet of the house can utilize the home water system and garden hose for completely extinguishing the ashy remains of the piles. If necessary, a fire line scraped down to the mineral soil with a garden rake, shovel, mattock, or pulaski can stop a fire creeping through the duff on the forest floor.

A 5-gallon firefighter's backpack water pump with spray nozzle is useful to keep slash pile fires from creeping too far. The backpack pump also can be filled quickly and put to use stopping a potential wildfire while it is very small. You can turn a three-quarter-ton or 1-ton pickup truck into a 200- or 350-gallon pumper for about $500 by installing a fiberglass water tank made for pickup trucks, a small gasoline pump, and a garden hose and fittings. If your home forest has reasonably good access by roads or truck trails, you will have your own fire engine available when and where you want it. The pumper truck can draw water from a pond or other reservoir through a 2- or 2.5-inch intake hose to fill the tank in five minutes or less. The empty fiberglass tank is light enough to be slipped on and off the truck by one person.

A simple and effective way to use fire around a homesite is to burn accumulations of slash and dead wood in piles. Some forest landowners carry out controlled burning in the light fuels on the forest floor around their homes, but this step requires more knowledge and often help from a rural fire department (Arno and Harrington 1995). Some farm and ranch insurance may cover customary burning associated with a landowner's farm and forestry operations, but it is important to check carefully with your insurer. Prescribed burning specialists are available in some areas to carry out treatments on forest lands. Prescribed burning on privately owned forest lands is not widespread in the West. In contrast, extensive burning is done in privately owned forests in the Southeast, backed up in some states by "right to burn" legislation.

Burning can be a useful part of forest restoration, but it is not essential. You can manage fuels without burning. If you are accustomed to or interested in using fire, it is important to learn safe procedures, check out insurance and liability considerations, consult with the local fire department, have sufficient helpers, tools, and equipment on hand, and proceed cautiously and responsibly.

Rewards for the Homeowner

From a scientific point of view, the effects of even a high-intensity fire are a fascinating illustration of the processes of disturbance and regeneration that shaped wildland forests. But from a human perspective, a forest where homes have burned is always a sad and sobering place. In most cases this loss is even more tragic because it is preventable. Adapting to life in the woods involves learning skills and knowledge previously in the exclusive realm of professional foresters and ecologists. But as insurance companies and fire suppression agencies change their policies to shift responsibility to individual homeowners for protecting homes and their surrounding forest from fire, forestry skills are no longer the province of professionals or serious hobbyists. It will take time, hard work, and yes, expense, to manage fuels on your woodlot. But if you are like many other forest homeowners who have done so, you'll also probably discover that your restored woodlot is even more attractive and "natural" looking than if you had taken no action. At the very least, you'll certainly have far more peace of mind when wildfire season rolls around every summer.

Chapter 12

Lessons from Nature: Will We Learn?

We westerners love our forests. National forests and national parks in the western United States garner more than 200 million visits annually with more coming each year. Two to three million of us now live in or adjacent to forests, most having migrated to the woods since 1970, the year of the first Earth Day. We have become forest residents at considerable expense and inconvenience. Now, we learn that many of our forests are in poor condition and even threaten our safety, all because we disrupted the vital natural fire process that renews western forests, and we have not adapted our forest dwellings and recreational facilities to fit into the fire-prone environment.

The Choice of Fire Exclusion

A century ago we almost chose to accommodate the fire-dependent character of western forests. Many observant and prominent people familiar with these forests argued for light burning in the lower elevations where they had frequently seen fire since the earliest explorations and settlement (Hoxie 1910, Powell 1891, Pyne 1982, 2001a). Even Gifford Pinchot, the founder of the U.S. Forest Service, at one time urged careful study of the ecological relationships of fire in our forests (Pinchot 1899). But instead of embracing the crucial role of fire in forests, we listened to another faction of well-meaning foresters, conservationists, and politicians who crusaded to eliminate fire. It has taken nearly a century for the implications of that fateful choice to be made manifest. Ninety years later, most forest biologists, land managers, and even government leaders now realize that our belief we could control nature was misplaced. With the decline of technocratic optimism and the rise of ecological consciousness, we see that the campaign against fire was ill conceived because it profoundly disrupts the ecological processes that produced the very forests fire suppression advocates sought to nurture.

By now conditions in forests that historically experienced frequent fires have changed so much that we no longer can simply use light burning to maintain them. These forests have grown thick with small trees, other fuels, and new homes. Today, smoke regulations and a myriad of other rules aim to protect specific features of the environment but actually hamper restoration of intact, self-sustaining ecosystems. Public mistrust of forest managers and entrenched habits within land management agencies prevent action. These impediments threaten to stifle the program to reduce fuels and restore fire in western forests that a politically divided White House and Congress nonetheless both endorsed in the fall of 2000.

A Chance to Reconsider

We are now confronted with an updated version of the historical debate over accepting fire in our forests. Given vast areas of fire-dependent forests that we want to maintain and protect, should we continue our attempts to keep fire out, or should we adopt widespread use of prescribed fire and other restoration treatments? In spite of growing support for changing course and restoring a semblance of the natural fire process, the failed crusade against fire continues to prove irresistible. The federal government responded to the record-setting 2000 wildfire season by increasing funding by $1.8 billion (USDA Forest Service 2001). The majority of this funding is for firefighting forces and equipment. However, some of the money is for an increase in forest restoration treatments such as prescribed burning, thinning, and other fuel reduction. Both of these program expansions can be justified. However, spending more to bolster fire suppression merely deals with a symptom of our inadequate program of fuels management. It does not treat the underlying cause of the problem. Unless we do large-scale prescribed burning and other fuel treatments in high-priority areas, increasing suppression efforts will just plow more money into an unwinnable war against fire (GAO 1999).

The old saw, "if all you have is a hammer, everything looks like a nail," comes to mind in the face of a continued institutional culture within the agencies responsible for fire management. In spite of new recommendations for fuels management and prescribed burning, all-out suppression remains the blunt instrument of choice. It is what everyone, the public included, is used to relying on, whether it works or not. Consider the campaign against the 217,000-acre Clear Creek Fire on the Salmon-Challis National Forest in central Idaho during the summer of 2000. Although the fire burned mostly in undeveloped mountainous terrain, it was near enough to populated valleys, including the town of Salmon, to inspire frenzied efforts to stop the advancing flames with heavy equipment. This included a 200-foot-wide bulldozed firebreak, not ultimately used for fire suppression because the fire did not reach it (Barker 2000). Despite the monumental expenditure of $71 million and the construction of more than 200 miles of bulldozed firelines, only rain and light snowfall finally quelled the blaze. These huge firelines would probably have proven ineffective anyhow since firebrands were igniting new blazes hundreds of yards beyond the existing fire. This war on a wildfire left an abundance of bulldozed soil suitable for invasion by noxious weeds, their seeds often inadvertently introduced on the heavy equipment rushed to the area during the heat of battle. In contrast, the 182,000-acre Wilderness Complex fire farther west in similar terrain was monitored and allowed to burn with very little suppression effort at a cost of about half a million dollars and with minimal environmental damage (Barker 2000).

In spite of our current ecological knowledge and technical expertise, increasing the area of prescribed burning and other restoration treatments is still politically a hard sell. It will require visionary leadership and commitment by all parties to work toward a common goal based on a broadly agreed upon definition of healthy and sustainable forests. In contrast, increasing firefighting efforts will not require any new agreement or public consent and will remain a routine task of government.

Perhaps the greatest resistance to forest restoration treatments comes from people who object to logging, especially on public lands, and believe that restoration is merely an excuse to continue conventional logging (Easthouse 2000). There is certainly cause for this concern. Timber management and logging were the highest-priority uses on many federal forestlands in the West for many years, and logging practices frequently damaged the land or streams with excessive road building, clearcutting, or scarification of the ground with bulldozers (Hirt 2000). Selective cutting often emphasized quickly removing the valuable trees without enough concern for the future forest. This emphasis on "getting out the cut" was tied to the postwar economic rebuilding effort initiated in the 1950s, and it continued through the 1980s. In the 1960s, ecologists and conservationists began to argue that these logging practices harmed the long-term health and recreational values of the forest. The growing environmental movement led to most of our modern environmental protection laws, including the National Environmental Protection Act, Clean Air Act, Clean Water Act, Endangered Species Act, and National Forest Management Act. It is worth remembering that at the time, support for environmental protection crossed ideological lines so strongly that President Nixon signed much of this legislation into law.

By the late 1990s, environmental interests making use of public appeal processes and court challenges and mustering public opinion had largely shut down the timber sale program on most federal forestlands. This stopped most abusive logging, but it also prevented an increasing number of well-designed, ecologically based forestry treatments that included carefully planned and administered tree harvesting. Unlike traditional logging, these restoration treatments rely on low-impact equipment and techniques. Unfortunately, the infamous Timber Salvage Rider passed by Congress in 1995 and ongoing battles to protect specific roadless areas and old growth have continued to undermine the Forest Service's credibility with environmental watchdogs. This mistrust of the agency has slowed acceptance of much-needed restoration initiatives.

The involvement of environmentally concerned citizens is critical to the success of restoration efforts. Without strong demands for restoration, fire exclusion will continue, and we will sink deeper into the quagmire of war against fire. Public opinion is clearly in transition and now is the time to make new choices and embrace new directions. On one hand, most people agree that the century-old campaign to rid western forests of fire has proven to be short sighted and has brought many problems to wild-land forests. On the other hand, proposals for using fire ecology as a basis for cutting and burning treatments to restore these forests are still met with skepticism. Critics most often object that wildfires will still burn through treated areas and that severe fires are natural and cannot be prevented by forest management. We have argued that areas with properly designed fuel treatments will moderate the intensity of a wildfire, greatly increasing the chance that trees will survive and also creating a zone where fire suppression can be more effective.

The second objection is harder to answer because it is in part based on a philosophical conviction that human intervention in natural systems is wrong or undesirable. But over 200 fire history studies show that there are contrasting natural fire regimes associated with different forest types. Some forests commonly experienced severe fires; others did not. Forest scientists consider the 1988 Yellowstone area fires to be a more-or-less natural event. In contrast, large, stand-replacement fires in the ponderosa pine and ponderosa pine–fir forests throughout much of the West in the last few decades are more severe and therefore more destructive than fires of past centuries.

The 1988 Yellowstone fires have been invoked to support the view that all forest management is ineffective in reducing fire severity. Critics claim that clearcuts with young trees along the national park's western boundary burned just as severely as the old lodgepole pine forest in the park. However, research indicates that lethal crown damage occurred in 89 percent of the survey plots in the old forest compared with only 20 percent of the plots in the adjacent treated stand, which consisted of saplings ranging from 1 to 16 feet tall (Omi and Kalabokidis 1991). Severe fires may be natural events that are quite acceptable in some forests. Management on a local scale can facilitate natural burning by allowing a resort or community to escape damage. A forested fuelbreak is certainly a more appropriate treatment for this situation than a clearcut, but the effectiveness of fuel reduction is similar.

Some people argue that forests with disrupted natural fire regimes can recover if we just allow lightning fires to burn. However, in forests that were historically in understory or mixed fire regimes, a high proportion of wildfires today are stand-replacing (Quigley, Haynes, and Graham 1996). These fires kill the old growth fire-dependent tree species that used to survive underburns. Moreover, with the thousands of homes at risk to wildfire in any given area of the West, governments simply will not allow fires to burn and potentially threaten homes or even entire communities. The only feasible way to allow natural fires to burn, even in large wilderness areas, is by carrying out fuel reduction treatments in a band around the area, particularly in the vicinity of homes and other developments (Omi 1996).

If We Stick with Fire Exclusion

Despite the logic and scientific evidence pointing to implementing fuels management, returning some semblance of natural fire processes remains a huge challenge. Is it worth the effort? What will happen if we just continue down the well-worn path of fire exclusion with minimal attention to controlling fuels? Recent trends suggest we follow this course at our peril.

WILDFIRE EMERGENCIES

Without fuel reduction, we will see more frequent wildfire emergencies resulting in more massive air pollution episodes in western communities and major economic losses to property owners and communities as facilities are damaged and recreation and tourism decline. Ever larger numbers of homeowners will be evacuated, and hundreds or perhaps thousands of homes will be lost during some wildfires. Homeowners will file claims and sue fire suppression agencies for losses allegedly the result of faulty suppression tactics. If we stay the course, the federal government will continue to be pressured to provide free fire protection, and there will be little incentive for homeowners to take actions needed to make their homes fire-resistant. Already, the federal government is reimbursing about 200 Los Alamos homeowners because their residences were burned by what has been deemed to be an improperly executed prescribed fire

(American Forests 2000). However, their losses could have been avoided if they had removed pine needle litter and other combustibles around their residences (Cohen 2000, Meeks 2001).

If we continue to rely on fire exclusion and ignore fuel buildup, fire-fighting itself may become more constrained because of increasing numbers of costly disputes over liability for fire damage. Several homeowners who were burned out during the 2000 wildfires in one canyon in western Montana had previously been unable to purchase fire insurance because of the high hazard of homes in dense forest, on steep terrain, and with poor access. However, the homeowners claim that a backfire set by federal firefighters burned their homes and that the federal government should therefore compensate them. Backburning is a common tactic in which a fire is set along the inside edge of a fire-control line in an area with light fuels. The backfire burns toward the oncoming wildfire and deprives it of fuel. However, it is risky to light a backfire in a dense forest. The winds may shift and push the fire in the wrong direction. If fuel reduction zones such as forested fuelbreaks existed in the vicinity of the homes, backburning would be much safer. As it is, homeowners living in heavy fuels face certain danger no matter the ignition source.

Because of ever larger numbers of homes at high risk from wildfire damage, firefighting efforts will be forced to concentrate further on protecting homes. Fewer resources will be available to control the actual spread of fire. This seems logical when fires are increasingly larger and more dangerous to approach, and firefighter safety is paramount. Still, without a large-scale program to reduce hazardous fuels, even relatively small, unimposing wildfires are a threat to flare up and overrun fire crews, as was the case with the 14 firefighters killed at Colorado's South Canyon Fire in 1994 and the 4 who died in July 2001 when a 10-acre fire in central Washington's Okanogan National Forest suddenly blew up after it was considered to be completely under control.

But not everyone shares the assumption that forest homeowners are entitled to heroic protection at the expense of all taxpayers. Australian firefighters helping with suppression efforts in the northern Rockies in 2000 were surprised that so much attention was being paid to saving private homes. In the wildlands of Australia, homeowners are responsible for providing their own wildfire protection. The Australian "Prepare, Stay, and Survive" program encourages homeowners to reduce fuels on and around their home and to assist in suppression efforts (Mutch 2001). A comparable philosophy of joining with others is traditional in American rural areas as well, where it resulted in the formation of volunteer fire departments. If the homeowner has to assume responsibility for fire protection, there will be an incentive to create fire-safe conditions at and around the home. This would be reinforced by standards set by insurers before they agree to provide coverage. Refusal of coverage by an insurer should be taken as a sign that the situation is too risky for anyone but the individual to assume responsibility.

Although federal fire policy has attempted to limit the government's role as the protector of homes from forest fires, political pressure during wildfire emergencies has induced federal suppression forces to provide expensive crews and equipment to protect residences. In the summer of 2000, one or more federally financed engine crews often were stationed at each house or vacation cabin in a threatened area, and the government provided and installed water pumps and sprinkler systems to protect the residence. With continuing building in hazardous forests, the practice of providing free protection funded by all U.S. taxpayers fails to induce people to take responsible action to safeguard their own homes.

ENVIRONMENTAL DAMAGE

Although western forests are well adapted to their historical fire regimes, forest values can be damaged as a result of our attempts to exclude fire. If we continue to exclude fire, magnificent forests of old growth trees and undergrowth plants characteristic of understory and mixed fire regimes will die out. Ponderosa pine, Jeffrey pine, sugar pine, western white pine, whitebark pine, western larch, and other long-lived trees will lose out in competition from shade-tolerant conifers. An increasing number of stand-replacement wildfires will also kill the remaining old trees and prevent development of future old growth (Fig. 12.1). Hundreds of thousands of miles of streamside, or riparian, forests historically associated with understory or mixed fire regimes will continue to be protected from all fire and fuel treatments, and will eventually burn in severe wildfires

Figure 12.1. This old growth ponderosa pine survived numerous fires prior to 1900 but was killed by a severe wildfire in 1988. Note the small firs in the foreground that crowned, causing the old pine to be severely scorched and to exhibit dead "frozen foliage." (USDA Forest Service photo.)

(Agee 1998). Fragmented populations of at-risk bull trout, salmon, and other native fish will be damaged or destroyed. Biological diversity of natural forest communities, including pioneer species of herbs, shrubs, and trees and the wildlife dependent upon them, will be degraded as historical mixed fire regimes continue to be replaced by stand-replacement regimes (M. K. Arno 1996, Olson 2000, Quigley, Haynes, and Graham 1996, Wheeler, Redman, and Tewksbury 1997).

If we continue to rely on fire exclusion in western forests, logging interests and environmental activists will persist in blaming each other for severe wildfires, and there will be little attention focused on what we could do to together to restore and perpetuate healthy wildland forests (Kiester 1999, Little 1998, 2000). Hundreds of miles of firelines will be bulldozed hurriedly in largely futile attempts to halt threatening wildfires. The firelines will be points of entry for noxious weeds and will cause erosion in some areas. To reduce erosion, slash from cleared trees will continue to be piled atop the firelines, but this will negate any value as a firebreak in the future by creating strips of untreated slash extending across the landscape. More effective and long-lasting fire barriers in the form of forested fuelbreaks and area-wide fuels management zones will not be constructed because they require planning and environmental review, and funding for their execution is still not readily available.

We do not mean to imply that fire suppression plays a trivial or irrelevant role in the successful management of fire and fuels. However, it is folly to rely on fire suppression alone and to largely ignore the need to employ other essential tools such as prescribed fire, thinning, slash disposal, other fuel treatments, and making forest homes fire resistant. Like most wars, a war on wildfire is very expensive and wasteful. We might just as soon declare "war on hurricanes" or some other powerful force of nature. Emergency firefighting funds are not directed at a useful task such as treating fuels to prevent damaging fire or accomplishing other steward-ship goals for the forest. Effective and ecologically sensitive firefighting depends on knowledge that the decision maker in charge of a suppression operation may not possess. Large firefighting campaigns rapidly assemble people and crews from many different agencies and jurisdictions all over the country, and often from Canada, Mexico, and Australia as well. Many of these crews and even the incident command teams in charge of a given wildfire campaign have little familiarity with the country, its terrain, weather, fuels, or associated environmental concerns. Local biologists or other natural resource specialists are supposed to be assigned to help interpret local conditions, but during the heat of a wildfire campaign this linkage to the land may be inadequate or unavailable.

Wildfire suppression is funded as a massive emergency operation. There is little time to weigh costs and benefits of different strategies, let alone the associated environmental effects. In contrast, prescribed burn-ing is conducted by locally experienced specialists who take time to plan all aspects of their operation. Prescribed fire is planned to achieve speci-fied, beneficial results.

The present management paradigm in many western forests on private and on public lands does little to restore and maintain forest structures that can reduce danger of severe fire damage. Instead, the emphasis is on either harvesting large trees or preserving dense forest growth. Continu-ing this course, we can expect to burn up a lot of trees that might otherwise develop into old growth or be useful for forest products. Even where the burned trees can be salvaged, they decay rapidly and are worth less than harvested live trees. Moreover, burned trees, unlike live trees, cannot be harvested when most needed by fluctuating timber markets, and post-fire salvage operations are just as controversial as other proposals to log public forests.

As long as we continue to let forests accumulate fuels and then burn while others are managed primarily to extract timber, we are failing to be good forest stewards. We will spend more and more on forests, but with even less success in maintaining natural biodiversity and the many values that people place on forests (Dekker-Robertson and Libby 1998, Foster 2000).

If We Choose Responsible Action

In spite of political obstacles and entrenched agency cultures, a new picture of fire-dependent forests in the West is emerging slowly but surely. Professional ecologists have focused attention on the vital importance of natural disturbances for maintaining the diversity of vegetation and wildlife habitat (Botkin 1990, Foster, Knight, and Franklin 1998). Private conservation organizations like The Nature Conservancy, the Rocky Mountain Elk Foundation, and land trusts located throughout much of the West increasingly recognize that preservation with continued exclusion of fire is often inappropriate for maintaining ecological values (Dickmann and Rollinger 1998). They have embraced prescribed burning and restoration forestry to maintain ecological values over the long term (Brown, Reinhardt, and Kramer 2001, Dickmann and Rollinger 1998, Reid 1998, Wilkinson 2001). They also often provide matching funds to help finance prescribed burning for ecological purposes on public lands. Their leadership amply demonstrates the feasibility, safety, and ecological benefits of restoration treatments. Many individual private landowners including graduates of forest stewardship courses are carrying out restoration treatments (Arno and Harrington 1995, Manning 2001) in a tradition reminiscent of T. B. Walker and other early twentieth century forest stewards in northern California (Hoxie 1910, Pyne 2001a).

If we cannot reach political consensus around the ecological condition

Figure 12.2. This prescribed burn was conducted into the night to take advantage of cooler conditions. The stand was commercially thinned, leaving the largest and most vigorous trees; some of the slash was removed and the remainder was burned to help create a suitable seedbed for ponderosa pine seedlings. (USDA Forest Service photo.)

of the forest, there is a greater chance that politicians will respond to the threat of property loss. Fuel reduction and management are urgently needed in strategically located zones near residential forests. Creation of forested fuelbreaks would allow broader use of natural fires or prescribed fire in more remote forests. Fuels management necessarily involves people in local community efforts to plan ecologically based practices to improve forest conditions (Brown, Reinhardt, and Kramer 2001, Gerritsma 2000, Kiester 1999, Little 1998, 2000). Fuels management or forest restoration could also employ many people to accomplish treatments on the ground (Fig. 12.2). Some of this work could be accomplished at low cost by mobilizing the thousands of wildland firefighters who have little to do during periods of slack wildfire activity.

Successful restoration treatments are designed to be appropriate for the forest type involved. In forests historically shaped by understory or mixed fire regimes, restoration treatments generally favor retention of vigorous and large trees. Restoration would lead toward forests patterned on presettlement conditions that have low hazard to wildfire and insect and disease epidemics. These treatments focus on maintaining desirable tree growth and understory vegetation, wildlife habitat, clean water, and sustainable timber harvest rather than maximizing income.

Still, large-scale restoration forestry on federal lands in the West could be largely self-financing (Fiedler and others 2001). Restoration practices can provide a broad range of useful wood products as well as firewood and fuel for generating electricity. This has important ecological and economic implications for the United States, which imports a large portion of the wood and energy it consumes. For instance, there is a growing need for environmentally friendly methods of generating electric power. Chipping some of the excess accumulations of small trees in western forests may now be economically attractive as a fuel source and could provide a fuel having less impact on air quality or global warming than fossil fuels. Using trees cut during restoration treatments can reduce our demand for imported wood from countries where logging is done with little concern for environmental impacts (Fiedler and others 2001, Foster 2000).

At the same time conservation groups such as the Sierra Club and the Native Forest Network have embraced the findings of fire ecology and some form of restoration forestry, they oppose deriving any economic gain from materials removed during restoration treatments (Sierra Club 2001, Native Forest Network 2001). Many environmentalists do not trust the Forest Service or other forestry agencies to oversee scientifically sound restoration and worry that creating a market for chips and small trees will lead to overharvesting in the guise of restoration, just as industrial logging overharvested mature trees sometimes in the name of "salvage" and "forest health." While we cannot claim that this will not happen, we feel that it is better for restoration to begin now by at least partially paying for itself, rather than waiting for massive congressional funding to subsidize all restoration efforts. To that extent, we are in agreement with *what* should be done, but differ as to *how* it might be accomplished. We hope to have contributed insight into the problem and offered possible solutions, but have no illusions that the debate is anywhere near to being settled.

Throughout this book, we have argued that we should look more closely to nature for cues to how we should manage wildland forests. Both we westerners and our forests will benefit from a revised perception of fire that acknowledges it as an inevitable and essential process. Rather than fighting fire as an implacable enemy, we should actively manage it in order to enjoy a healthy and sustainable wildland forest. The century-old fire suppression policy was based on a forestry paradigm transplanted from Europe; it ignored contradictory research and experience and promoted the political aspirations of the fledgling U.S. Forest Service. Today we have fine-grained empirical pictures and robust theoretical models of forest structure and fire effects that much more clearly discern the differences between historical and current conditions. It only seems reasonable that the policies affecting millions of acres of public lands be based on our current knowledge and social values. If we act on the knowledge and expertise we now possess, the forests of the twenty-first century will more closely resemble those of the mid nineteenth century than anything most of us have personally seen. We will never be entirely free of fire's demands, but we still have time to heed the advice of California timberman George Hoxie (Hoxie 1910). He argued long ago, and we echo his sentiment today, that we had best adopt fire as our servant. Else it surely will be our master.

Appendix

Getting Help: Information and Educational Resources for Forest Landowners

State Service Foresters

State service foresters provide management assistance and fire protection for much of the nation's private forestland. Private landowners interested in learning more about natural resource protection and educational programs such as forest stewardship can contact their service forester (SF) at the address below or explore the website: www.stateforesters.org.

Arizona—SF, State Land Department, 1616 W. Adams, Phoenix AZ 85007; 602/542-2515

California—SF, Dept. of Forestry and Fire Protection, PO Box 944246, Sacramento, CA 94244; 916/653-7772

Colorado—SF, State Forest Service, Colorado State University, 203 Forestry Bldg., Fort Collins, CO 80523; 970/491-6303

Idaho—SF, Department of Lands, PO Box 83720, Boise, ID 83720; 208/334-0200

Montana—SF, Department of Natural Resources and Conservation,

Forestry Division, 2705 Spurgin Road, Missoula, MT 59804; 406/542-4300

Nevada—SF, Division of Forestry, 1201 Johnson, Suite D, Carson City, NV 89706; 702/684-2501

New Mexico—SF, Forestry Division, PO Box 1948, Sante Fe, NM 87504; 505/476-3328

Oregon—SF, Department of Forestry, 2600 State St., Salem, OR 97310; 503/945-7211

South Dakota—SF, Resource Conservation and Forestry, 523 E. Capitol Ave., Pierre, SD 57501; 605/773-3623

Utah—SF, Department of Natural Resources, 1594 W. North Temple, Suite 3520, Salt Lake City, UT 84114; 801/538-5530

Washington—SF, Department of Natural Resources, PO Box 47037, Olympia, WA 98504; 360/902-1011

Wyoming—SF, Forestry Division, 1100 West 22nd St., Cheyenne, WY 82002; 307/777-7586

Extension Forestry

The U.S. Department of Agriculture and state agricultural colleges work together to develop educational materials, demonstrations, and workshops to provide landowners with information on forest ecology and natural resource management and protection. State Extension Forestry (EF) offices are listed below:

Arizona—EF, University of Arizona, PO Box 210043, Tucson, AZ 85721; 520/621-7257

California—EF, University of California, Department of Environmental Science, 163 Mulford Hall, Berkeley, CA 94720; 510/643-5428

Colorado—EF, Colorado State University, Department of Forest Science, 100 Natural Resources Lab, Fort Collins, CO 80523; 970/491-7780

Idaho—EF, University of Idaho, College of Forestry, Moscow, ID 83844; 208/885-6356

Montana—EF, University of Montana, School of Forestry, Missoula, MT 59812; 406/243-2773

Nevada—EF, University of Nevada, Environmental Resource Sciences, Reno, NV 89512; 702/784-4039

New Mexico—EF, New Mexico State University, Box 3AE, Las Cruces, NM 88003; 505/646-2053

Oregon—EF, Oregon State University, College of Forestry, 109 Richardson Hall, Corvallis, OR 97331; 541/737-3700

South Dakota—EF, South Dakota State University, Box 2140A, Brookings, SD 57007; 605/688-4737

Utah—EF, Utah State University, Department of Forest Resources, Logan, UT 84322; 435/797-4056

Washington—EF, Washington State University, Department of Natural Resource Sciences, PO Box 646410, Pullman, WA 99164; 509/335-2964

Wyoming—EF, University of Wyoming, Department of Renewable Resources, PO Box 3354, Laramie, WY 82071; 307/766-5130

The Tree Farm System

The American Tree Farm System provides conservation education and certification to promote sustainable forest management on private lands.

It is a nonprofit, nongovernmental organization established in 1941 with programs in each state. Current information on how to contact your state tree farm program is on the website: www.treefarmsystem.org. It can also be obtained by phoning the American Forest Foundation at 1-888/889-4466.

Woodland Owner Associations

Woodland owner associations provide an educational forum and represent the interests of small woodland owners so they will be able to practice good stewardship. The following western states have such associations.

California—Forest Landowners of California, 980 Ninth Street, Suite 1600, Sacramento, CA 95814; 916/972-0273

Colorado—Colorado Forestry Association, PO Box 270132, Fort Collins, CO 80527; 970/491-6303

Idaho—Idaho Forest Owners Association, PO Box 1257, Coeur d'Alene, ID 83816; 208/762-9303

Montana—Montana Forest Owners Association, 17975 Ryans Lane, Evaro, MT 59802; 406/726-3787

New Mexico—Forest Trust, PO Box 519, Sante Fe, NM 87504; 505/983-8992

Oregon—Oregon Small Woodlands Association, 1775 32nd Place NE, Suite C, Salem, OR 97302; 503/588-1813

Utah—Utah Woodland Owners Council, 2829 Sleep Hollow Drive, Salt Lake City, UT 84117; 801/277-1615

Washington—Washington Farm Forestry Association, 110 W. 26th Avenue, Olympia, WA 98507; 360/459-0984

Firewise Program

Firewise is an educational program sponsored by the National Association of State Foresters that provides information for homeowners who live in or adjacent to wildland fuels. Firewise works with communities to establish workshops and other events that raise awareness of wildfire risk and of alternatives that lower it. The website is: www.firewise.org.

References

Abrams, M. D. 1992. Fire and the development of oak forests. *Bioscience* 42(5):346–353.

Agee, J. K. 1993. *Fire ecology of Pacific Northwest forests.* Washington, D.C.: Island Press.

———. 1996. Fire in restoration of Oregon white oak woodlands. Pages 72–73 in C. C. Hardy and S. F. Arno, eds., *The use of fire in forest restoration.* USDA Forest Service Intermountain Research Station, General Technical Report INT-GTR-341, Ogden, Utah.

———. 1997. The severe weather wildfire—too hot to handle? *Northwest Science* 71(1):153–156.

———. 1998. The landscape ecology of western forest fire regimes. *Northwest Science* 72(Special Issue):24–34.

———. 2000. Wilderness fire science: A state-of-knowledge review. Pages 5–22 in D. N. Cole, S. F. McCool, W. T. Borrie, and J. O'Loughlin, comps., *Wilderness science in a time of change conference.* Vol. 5, *Wilderness ecosystems, threats, and management.* USDA Forest Service, Rocky Mountain Research Station, RMRS-P-15-Vol. 5, Ogden, Utah.

———. 2001. Personal communication. Comments written to author in review of draft manuscript.

Agee, J. K., B. Bahro, M. A. Finney, P. N. Omi, D. B. Sapsis, C. N. Skinner, J. W. van Wagtendonk, and C. P. Weatherspoon. 2000. The use of shaded fuelbreaks in landscape fire management. *Forest Ecology and Management* 127:55–66.

American Forests. 1995. *Forest health and fire danger in inland Western forests.* Proceedings of American Forests Conference, Sept. 8–9, 1994, Spokane, Washington. Washington, D.C.: Harman Press.

———. 2000. News from the world of trees: Lessons from Los Alamos. *American Forests* 106(3):13.

Anderson, H. E. 1968. Sundance fire: An analysis of fire phenomena. USDA

Forest Service, Intermountain Forest and Range Experiment Station, Research Paper INT-56, Ogden, Utah.

Anderson, H. E., and J. K. Brown. 1988. Fuel characteristics and fire behavior considerations in the wildlands. Pages 124–130 in W. C. Fischer and S. F. Arno, comps, *Protecting people and homes from wildfire in the Interior West: Proceedings of the symposium and workshop.* USDA Forest Service, Intermountain Research Station, General Technical Report INT-251, Ogden, Utah.

Anderson, J. E., M. Ellis, and W. H. Romme. 1998. Establishment, growth, and survival of lodgepole pine *(Pinus contorta)* on diverse sites burned in the Yellowstone fires of 1988. *Yellowstone Science* 6(2):25.

Anderson, L., C. E. Carlson, and R. H. Wakimoto. 1987. Forest fire frequency and western spruce budworm outbreaks in western Montana. *Forest Ecology and Management* 22:251–260.

Anderson, M. K., and M. J. Moratto. 1996. Native American land-use practices and ecological impacts. Pages 187–206 in *Sierra Nevada Ecosystem Project: Final report to Congress,* vol. 2. Davis: University of California, Centers for Water and Wildland Resources.

Arno, M. K. 1996. Reestablishing fire-adapted communities to riparian forests in the ponderosa pine zone. Pages 42–43 in C. C. Hardy and S. F. Arno, eds., *The use of fire in forest restoration.* USDA Forest Service, Intermountain Research Station, General Technical Report INT-341, Ogden, Utah.

Arno, S. F. 1976. The historical role of fire on the Bitterroot National Forest. USDA Forest Service, Intermoutain Forest and Range Experiment Station, Research Paper INT-187, Ogden, Utah.

———. 1986. Whitebark pine cone crops: A diminishing source of wildlife food. *Western Journal of Applied Forestry* 1:92–94.

———. 1999. Undergrowth response, Shelterwood Cutting Unit. Pages 36–37 in Y. H. Smith and S. F. Arno, eds., *Eighty-eight years of change in a managed ponderosa pine forest.* USDA Forest Service, Rocky Mountain Research Station, General Technical Report RMRS-GTR-23, Ogden, Utah.

———. 2000. Fire regimes in western forest ecosystems. Pages 97–120 in J. K. Brown, ed., *Effects of fire on flora.* USDA Forest Service, Rocky Montain Research Station, General Technical Report RM-42, vol. 2, Fort Collins, Colorado.

———. 2001. Community types and natural disturbance processes. Pages 74–88 in D. F. Tomback, S. F. Arno, and R. E. Keane, eds., *Whitebark pine communities ecology and restoration.* Washington, D.C.: Island Press.

Arno, S. F., and J. K. Brown. 1989. Managing fire in our forests: Time for a new initiative. *Journal of Forestry* 87(12):44–46.

Arno, S. F., and G. E. Gruell. 1983. Fire history at the forest–grassland ecotone in southwestern Montana. *Journal of Range Management* 36:332–336.

———. 1986. Douglas-fir encroachment into mountain grasslands in southwestern Montana. *Journal of Range Management* 39:272–275.

Arno, S. F., and M. G. Harrington. 1995. Use thinning and fire to improve forest health and wildlife habitat. *Tree Farmer* May/June:6–8, 23.

———. 1998. The Interior West: Managing fire-dependent forests by simulating natural disturbance regimes. Pages 53–62 in *Forest management into the next century: What will make it work?* Madison, Wis.: Forest Products Society.

Arno, S. F., and T. D. Petersen. 1983. *Variation in estimates of fire intervals: A closer look at fire history on the Bitterroot National Forest.* USDA Forest Service Intermountain Forest and Range Experiment Station, Research Paper INT-301, Ogden, Utah.

Arno, S. F., and K. M. Sneck. 1977. *A method for determining fire history in coniferous forests of the Mountain West.* USDA Forest Service Intermountain Forest and Range Experiment Station, General Technical Report INT-42, Ogden, Utah.

Arno, S. F., D. J. Parsons, and R. E. Keane. 2000. Mixed-severity fire regimes in the Northern Rocky Mountains: Consequences of fire exclusion and options for the future. Pages 225–232 in D. N. Cole, S. F. McCool, W. T. Borrie, and J. O'Loughlin, comps., *Wilderness science in a time of change conference.* Vol. 5, *Wilderness ecosystems, threats, and management.* USDA Forest Service, Rocky Mountain Research Station, RMRS-P-15-Vol. 5, Ogden, Utah.

Arno, S. F., E. D. Reinhardt, and J. H. Scott. 1993. *Forest structure and landscape patterns in the subalpine lodgepole pine type: A procedure for quantifying past and present conditions.* USDA Forest Service Intermountain Research Station, General Technical Report INT-294, Ogden, Utah.

Arno, S. F., J. H. Scott, and M. G. Hartwell. 1995. *Age-class structure of old growth ponderosa pine/Douglas-fir stands and its relationship to fire history.* USDA Forest Service Intermountain Research Station, Research Paper INT-481, Ogden, Utah.

Arno, S. F., D. G. Simmerman, and R. E. Keane. 1985. *Forest succession on four habitat types in western Montana.* USDA Forest Service Intermountain Forest and Range Experiment Station, General Technical Report INT-177, Ogden, Utah.

Arno, S. F., H. Y. Smith, and M. A. Krebs. 1997. *Old growth ponderosa pine and western larch stand structures: Influences of pre-1900 fires and fire exclusion.* USDA Forest Service Intermountain Research Station, Research Paper INT-495, Ogden, Utah.

Arno, S. F., M. G. Harrington, C. E. Fiedler, and C. E. Carlson. 1995. Restoring fire-dependent ponderosa pine forests in western Montana. *Restoration and Management Notes* 13(1):32–36.

Bailey, D. W., and J. Losensky. 1996. *Fire in western Montana ecosystems: A strategy for accomplishing ecosystem management through the effective use of prescribed fire in the Lolo National Forest.* Missoula, Mont.: Lolo National Forest.

Baisan, C. H., and T. W. Swetnam. 1997. *Interactions of fire regimes and land use in the central Rio Grande valley.* USDA Forest Service, Rocky Mountain Forest and Range Experiment Station, Research Paper RM-RP-330, Fort Collins, Colorado.

Barbouletos, C. S., L. Z. Morelan, and F. O. Carroll. 1998. We will not wait: Why prescribed fire must be implemented on the Boise National Forest. *Proceedings—Tall Timbers Fire Ecology Conference* 20:27–30. Tall Timbers Research Station, Tallahassee, Florida.

Barker, R. 2000. Fire officials weigh damage in wake of Clear Creek fire. *Idaho Statesman,* 22 September, 1A, 11A.

Barrett, S. W. 1981. Relationship of Indian-caused fires to the ecology of western Montana forests. M.S. thesis, University of Montana, Missoula, Montana.

———. 1982. *Fires' influence on ecosystems of the Clearwater National Forest: Cook Mountain fire history inventory.* USDA Forest Service, Clearwater National Forest, Fire Management, Office Report, Orofino, Idaho.

———. 1985. Living artifacts: Indian-scarred trees. *American Forests* 91(7):39.

———. 1994. Fire regimes on andesitic mountain terrain in northeastern Yellowstone National Park, Wyoming. *International Journal of Wildland Fire* 4:65–76.

Barrett, S. W., and S. F. Arno. 1982. Indian fires as an ecological influence in the Northern Rockies. *Journal of Forestry* 80:647–651.

———. 1988. *Increment-borer methods for determining fire history in coniferous forests.* USDA Forest Service Intermountain Research Station, General Technical Report INT-244, Ogden, Utah.

Barrett, S. W., S. F. Arno, and C. H. Key. 1991. Fire regimes of western larch–lodgepole pine forests in Glacier National Park, Montana. *Canadian Journal of Forestry Research* 21:1711–1720.

Barrett, S. W., S. F. Arno, and J. P. Menakis. 1997. *Fire episodes in the inland Northwest (1540–1940) based on fire history data.* USDA Forest Service Intermountain Research Station, General Technical Report INT-370, Ogden, Utah.

Becker, R. R., and T. S. Corse. 1997. The Flathead Indian Reservation: Resetting the clock with uneven-aged management. *Journal of Forestry* 95(11):29–32.

Benedict, G. W., L. R. Swan, and R. A. Belnap. 1991. Evolution and implementation of a fire management program which deals with high-intensity fires on the Payette National Forest in central Idaho. *Proceedings—Tall Timbers Fire Ecology Conference* 17:339–351. Tall Timbers Research Station, Tallahassee, Florida.

Bessie, W. C., and E. A. Johnson. 1995. The relative importance of fuels and weather on fire behavior in subalpine forests. *Ecology* 76:747–762.

Biondi, F. 1996. Decadal-scale dynamics at the Gus Pearson Natural Area: Evidence for inverse (a)symmetric competition? *Canadian Journal of Forestry Research* 26:1397–1406.

Biswell, H. H. 1989. Prescribed burning in California wildlands vegetation management. Berkeley: University of California Press.

Biswell, H. H., H. R. Kallander, R. Komarek, R. J. Vogl, and H. Weaver. 1973. *Ponderosa fire management: A task force evaluation of controlled burning in ponderosa pine forests of central Arizona.* Tall Timbers Research Station, Miscellaneous Publication 2, Tallahassee, Florida,

Bonnicksen, T. M., and E. C. Stone. 1981. The giant sequoia–mixed conifer forest community characterized through pattern analysis as a mosaic of aggregations. *Forest Ecology and Management* 3:307–328.

Botkin, D. B. 1990. *Discordant harmonies: A new ecology for the twenty-first century.* New York: Oxford University Press.

Boyd, R. 1986. Strategies of Indian burning in the Willamette Valley. *Canadian Journal of Anthropology* 5(1):65–86.

————, ed. 1999. *Indians, fire, and the land in the Pacific Northwest.* Corvallis: Oregon State University Press.

Bradley, A. F. 1984. Rhizome morphology, soil distribution, and the potential fire survival of eight woody understory species in western Montana. M.S. thesis, University of Montana, Missoula.

Brown, A. A. 1943. Comments. *Journal of Forestry* 41:14–15.

Brown, D. E., and N. B. Carmony. 1990. *Aldo Leopold's Wilderness: Selected early writings by the author of A Sand County Almanac.* Mechanicsburg, Pa.: Stackpole Books.

Brown, J. K. 1975. Fire cycles and community dynamics in lodgepole pine forests. Pages 429–456 in D. M. Baumgartner, ed., *Proceedings: Management of lodgepole pine ecosystems symposium.* Pullman: Cooperative Extension, Washington State University.

————. 1993. A case for management ignitions in wilderness. *Fire Management Notes* 53–54 (4):3–8.

————. 1995. Fire regimes and their relevance to ecosystem management. Pages 171–178 in *Proceedings of the Society of American Foresters National Convention.* Bethesda, Md.: Society of American Foresters.

————. 2000. Chapter 1: Introduction and fire regimes. Pages 1–8 in J. K. Brown, ed., *Wildland fire in ecosystems: Effects of fire on flora.* USDA Forest Service, Rocky Mountain Research Station, General Technical Report RM-42, vol. 2, Fort Collins, Colorado.

Brown, J. K., and E. D. Reinhardt. 1991. Predicting and managing fuel conditions in the Interior West. Pages 419–429 in P. L. Andrews and D. F. Potts, eds., *Proceedings of the 11th Conference on Fire and Forest Meteorology,* Missoula, Montana, April 16–19, 1991. Bethesda, Md.: Society of American Foresters.

Brown, J. K, E. D. Reinhardt, and K. A. Kramer. 2001. *Coarse woody debris and succession in the recovering forest.* USDA Forest Service, Fire Sciences Laboratory, Fire Effects Unit, contract report for the Bitterroot National Forest, Missoula, Montana.

Brown, J. K., S. F. Arno, S. W. Barrett, and J. P. Menakis. 1994. Comparing the prescribed natural fire program with presettlement fires in the Selway–Bitterroot Wilderness. *International Journal of Wildland Fire* 4:157–168.

Brown, J. K., S. F. Arno, L. S. Bradshaw, and J. P. Menakis. 1995. Comparing the Selway–Bitterroot fire program with presettlement fires. Pages 48–54 in J. K. Brown, R. W. Mutch, C. W. Spoon, and R. H. Wakimoto, technical coordinators, *Proceedings: Symposium on fire in wilderness and park management.* USDA Forest Service, General Technical Report, INT-320, Ogden, Utah.

Brown, P. M., and C. H. Sieg. 1996. Fire history in interior ponderosa pine communities of the Black Hills, South Dakota, USA. *International Journal of Wildland Fire* 6:97–105.

Brown, P. M., and T. W. Swetnam. 1994. A cross-dated fire history from coast redwood near Redwood National Park, California. *Canadian Journal of Forestry Research* 24:21–31.

Brown, P. M., D. R. D'Amico, A. T. Carpenter, and D. M. Andrews. 2001. Restoration of montane ponderosa pine forests in the Colorado Front Range: A forest ecosystem management plan for the City of Boulder, Colorado. *Ecological Restoration* 19(1):19–26.

Buchanan, J. B., and L. L. Irwin. 1993. Fire suppression and management of spotted owl habitat in the Wenatchee National Forest. *Journal of Raptor Research* 27(1):86–87.

Camp, A. E., P. F. Hessburg, and R. L. Everett. 1996. Dynamically incorporating late-successional forest in sustainable landscapes. Pages 20–23 in C. C. Hardy and S. F. Arno, eds., *The use of fire in forest restoration.* USDA Forest Service, Intermountain Research Station, General Technical Report INT-341, Ogden, Utah.

Carcaillet, C. 1998. A spatially precise study of Holocene fire history, climate, and human impact within the Maurienne valley, north French Alps. *Journal of Ecology* 86:384–396.

Chapman, H. H. 1926. *Factors determining natural reproduction of longleaf pine on cut-over lands in LaSalle Parish, Louisiana.* Yale University, School of Forestry, Bulletin No. 16. New Haven, Conn.

Christensen, N. L., J. K. Agee, P. F. Brussard, J. Hughes, D. H. Knight, G. W. Minshall, J. M. Peek, S. J. Pyne, F. J. Swanson, J. W. Thomas, S. Wells, S. E. Williams, and H. A. Wright. 1989. Interpreting the Yellowstone fires. *Bioscience* 39:678–685.

Clark, J. S., J. Merkt, and H. Muller. 1989. Post-glacial fire, vegetation, and human history on the northern alpine forelands, south-western Germany. *Journal of Ecology* 77:897–925.

Clements, F. E. 1910. *The life history of lodgepole burn forests.* USDA Forest Service, Bulletin 79. Washington, D.C.: GPO.

Cochran, P. H., and J. W. Barrett. 1998. *Thirty-five-year growth of thinned and unthinned ponderosa pine in the Methow Valley of northern Washington.* USDA Forest Service, Pacific Northwest Research Station, Research Paper PNW-502, Portland, Oregon.

Cohen, J. D. 2000. *Examination of the home destruction in Los Alamos associated with the Cerro Grande Fire.* USDA Forest Service, Rocky Mountain Research Station, Fire Sciences Laboratory, Fire Behavior Unit, Office Report, Missoula, Mont.

Cohen, S., and D. Miller. 1978. *The big burn: The Northwest's forest fire of 1910.*

Missoula, Mont.: Pictoral Histories Publication Company.

Coloff, S. G. 1995. Air quality and prescribed fire: Striving for a common goal. Pages 72–73 in J. K. Brown, R. W. Mutch, C. W. Spoon, and R. H. Wakimoto, technical coordinators, *Proceedings: Symposium on fire in wilderness and park management.* USDA Forest Service, Intermountain Research Station, General Technical Report INT-GTR-320, Ogden, Utah.

Cooper, C. F. 1960. Changes in vegetation, structure, and growth of southwestern pine forests since white settlement. *Ecological Monographs* 30(2):129–164.

Core, J. E., and J. L. Peterson 2001 (in press). State smoke management programs. In C. C. Hardy, R. D. Ottmar, J. L. Peterson, and others, eds. comps., *Smoke management guide for prescribed and wildland fire: 2001 edition.* National Interagency Fire Center, National Wildfire Coordinating Group, NWCG-PMS-420-2, Chapter 4.2, Boise, Idaho.

Covington, W. 1996. Implementing adaptive ecosystem restoration in western long-needled pine forests. Pages 44–48 in W. Covington and P. K. Wagner, technical coordinators, *Conference on adaptive ecosystem restoration and management.* USDA Forest Service, Rocky Mountain Forest and Range Experiment Station, General Technical Report RM-278, Fort Collins, Colorado.

Covington, W. W., and M. M. Moore. 1994a. Postsettlement changes in natural fire regimes and forest structure: Ecological restoration of old-growth ponderosa pine forests. *Journal of Sustainable Forestry* 2(1/2):153–182.

―――. 1994b. Southwestern ponderosa pine forest structure: Changes since Euro-American settlement. *Journal of Forestry* 92(1):39–47.

Covington, W. W., R. L. Everett, R. Steele, L. L. Irwin, T. A. Daer, and A. N. D. Auclair. 1994. Historical and anticipated changes in forest ecosystems of the Inland West of the United States. *Journal of Sustainable Forestry* 2(1):13–64.

Crane, M. F., J. Habeck, and W. C. Fischer. 1983. Early post-fire revegetation in a western Montana Douglas-fir forest. USDA Forest Service, Intermountain Forest and Range Experiment Station, General Technical Report INT-GTR-319, Ogden, Utah.

Cutter, B. E., and R. P. Guyette. 1994. Fire frequency on an oak–hickory ridgetop in the Missouri Ozarks. *American Midland Naturalist* 132:393–398.

Czech, B. 1996. Challenges to establishing and implementing sound natural fire policy. *Renewable Resources Journal* 14(2):14–19.

Daniels, O. L. 1991. A forest supervisor's perspective on the prescribed natural fire. *Proceedings—Tall Timbers Fire Ecology Conference* 17:361–366. Tall Timbers Research Station, Tallahassee, Florida.

Daubenmire, R. 1968. *Plant communities: A textbook of plant synecology.* New York: Harper and Row.

Davies, J. 1980. *Douglas of the forests: The North American journals of David Douglas.* Seattle: University of Washington Press.

Davis, K. P. 1959. *Forest fire control and use.* New York: McGraw-Hill.

DeByle, N. V., C. D. Bevins, and W. C. Fischer. 1987. Wildfire occurrence in aspen in the interior western United States. *Western Journal of Applied Forestry* 2(3):73–76.

DeByle, N. V., and R. P. Winokur, eds. 1985. *Aspen: Ecology and management in the western United States.* USDA Forest Service, Rocky Mountain Forest and Range Experiment Station, General Technical Report RM-119, Fort Collins, Colorado.

Dekker-Robertson, D. L., and W. J. Libby. 1998. American forest policy: Global ethical tradeoffs. *Bioscience* 48(6):471–477.

Dell'Orto, G. 2001. Turning back time: Interior, Forest Service leaders tour Arizona project to restore forest health. *Missoulian,* 9 August, A-2.

DeLuca, T. H. 2000. Soils and nutrient considerations. Pages 23–25 in H. Y. Smith, ed., *Proceedings: The Bitterroot Ecosystem Management Research Project: What we have learned symposium.* USDA Forest Service, Rocky Mountain Research Station RMRS-P-17, Ogden, Utah.

Despain, D. G. 1990. *Yellowstone vegetation.* Boulder, Colorado: Roberts Rinehart Publishing.

Despain, D. G., and W. H. Romme. 1991. Ecology and management of high-intensity fires in Yellowstone National Park. *Proceedings—Tall Timbers Fire Ecology Conference* 17:43–58. Tall Timbers Research Station, Tallahassee, Florida.

Devlin, S. 1995. Fighting wildfires. Montana Business Annual. *Montana Magazine,* special supplement, January/February, 26–31.

———. 2000. Check with reality: Intense blazes of 2000 may be wake-up call to return fire to forests. *Missoulian,* 22 August, A1, A6–7.

———. 2001a. Clear picture of filthy air: Study provides comprehensive analysis of pollution caused by wildfires of 2000. *Missoulian,* 4 February, A1, A12.

———. 2001b. Deal protects 53,000 acres of riverfront timberland. *Missoulian,* 15 June, A1, A4.

Dickmann, D. I., and J. L. Rollinger. 1998. Fire for restoration of communities and ecosystems. *Bulletin of the Ecological Society of America* 79(2):157–160.

Duchesne, L. C., and B. C. Hawkes. 2000. Fire in northern ecosystems. Pages 35–52 in J. K. Brown and J. K. Smith, eds., *Wildfire in ecosystems: Effects of fire on flora.* USDA Forest Service, Rocky Mountain Research Station, General Technical Report RMRS-GTR-42-Vol. 2, Ogden, Utah.

Duncan, F. L. 1992. Botanical reflections of the Encuentro and the Contact Period in southern Marin County, California. Ph.D. dissertation, University of Arizona, Tucson.

Easthouse, K. 2000. Drawing fire. *Forest Magazine,* November/December, 35–40.

The Economist. 2001. The blaze next time. 28 April, 32–33.

Eliot, W. A. 1948. *Forest trees of the Pacific Coast.* New York: G. P. Putnam's Sons.

Evans, J. W. 1990. *Powerful Rockey: The Blue Mountains and the Oregon Trail, 1811–1883.* Enterprise: Eastern Oregon State College, Pika Press.

Everett, R., D. Schellhaas, D. Spurbeck, P. Ohlson, D. Keenum, and T. Anderson. 1997. Structure of northern spotted owl nest stands and their historical conditions on the eastern slope of the Pacific Northwest Cascades, USA. *Forest Ecology and Management* 94:1–14.

Ferry, G. W., R. G. Clark, R. E. Montgomery, R. W. Mutch, W. P. Leenhouts, and G. T. Zimmerman. 1995. Altered fire regimes within fire-adapted ecosystems. Pages 222–224 in *Our living resources: A report to the nation on the distribution, abundance, and health of U.S. plants, animals, and ecosystems.* U.S. Department of the Interior, National Biological Survey, Washington, D.C.

Fiedler, C., and S. F. Arno. 2000. A conservation easement that perpetuates ecosystem-based management on a Montana ranch. Page 143 in H. Y. Smith, ed., *Proceedings: The Bitterroot Ecosystem Management Research Project: What have we learned symposium.* USDA Forest Service, Rocky Mountain Research Station, Proceedings RMRS-P-17, Ogden, Utah.

Fiedler, C. E. 2000a. Restoration treatments promote growth and reduce mortality of old-growth ponderosa pine (Montana). *Ecological Restoration* 18:117–119.

———. 2000b. Silvicultural treatments. Pages 19–20 in H. Y. Smith, ed., *Proceedings: The Bitterroot Ecosystem Management Research Project: What have we learned symposium.* USDA Forest Service, Rocky Mountain Research Station RMRS-P-17, Ogden, Utah.

Fiedler, C. E., and J. F. Cully. 1995. A silvicultural approach to develop Mexican spotted owl habitat in Southwest forests. *Western Journal of Applied Forestry* 10:144–148.

Fiedler, C. E., R. R. Becker, and S. A. Haglund. 1988. Preliminary guidelines for uneven-aged silvicultural prescriptions in ponderosa pine. Pages 235–241 in D. M. Baumgartner and J. E. Lotan, comps. and eds., *Ponderosa pine: The species and its management.* Pullman: Washington State University Cooperative Extension.

Fiedler, C. E., S. F. Arno, C. E. Keegan, and K. A. Blatner. 2001. Overcoming America's wood deficit: An overlooked option. *Bioscience* 51(1):53–58.

Fiedler, C. E., C. E. Keegan III, D. P. Wichman, and S. F. Arno. 1999. Product and economic implications of ecological restoration. *Forest Products Journal* 49(2):19–23.

Filip, G. M. 1994. Forest health decline in central Oregon: A 13-year case study. *Northwest Science* 68(4):233–240.

Finney, M. A. 2001. Design of regular landscape fuel teatment patterns for modifying fire growth and behavior. *Forest Science* 47(2):219–228.

Finney, M. A., and R. E. Martin. 1989. Fire history in a *Sequoia sempervirens* forest at Salt Point State Park, California. *Canadian Journal of Forestry Research* 19:1451–1457.

———. 1992. Short fire intervals recorded by redwoods at Annadel State Park, California. *Madrono* 39:251–262.

Fletcher, F. 1652. *The world encompassed by Sir Francis Drake* [collected out of the notes of Master Francis Fletcher]. Printed at London for Nicholas Bourne, dwelling at the South entrance of the royal Exchange.

Fonda, R. W., and L. C. Bliss. 1969. Forest vegetation of the montane and sub-alpine zones, Olympic Mountains, Washington. *Ecological Monographs* 39:271–301.

Foster, D. R. 2000. Conservation lessons and challenges from ecological history. *Forest History Today*, fall, 2–11.

Foster, D. R., D. H. Knight, and J. F. Franklin. 1998. Landscape patterns and legacies resulting from large, infrequent forest disturbances. *Ecosystems* 1:497–510.

Franklin, J. F., and C. T. Dyrness. 1973. *Natural vegetation of Oregon and Washington.* USDA Forest Service, Pacific Northwest Forest and Range Experiment Station, General Technical Report PNW-8, Portland, Oregon.

Franklin, J. F., D. R. Berg, D. A. Thornburgh, and J. C. Tappeiner. 1999. Alternative silvicultural approaches to timber harvesting: Variable retention harvest systems. Pages 111–139 in K. A. Kohm and J. F. Franklin, eds., *Creating a forestry for the 21st century.* Washington, D.C.: Island Press.

Fritz, E. 1931. The role of fire in the redwood region. *Journal of Forestry* 29:939–950.

Fuller, M. 1991. Forest fires: An introduction to wildland fire behavior, management, firefighting, and prevention. New York: John Wiley and Sons.

Gaines, W. L., R. A. Strand, and S. D. Piper. 1997. Effects of the Hatchery Complex Fires on northern spotted owls in the eastern Washington Cascades. Pages 123–129 in J. M. Greenlee, ed., *Proceedings: Conference on fire effects on rare and endangered species and habitats,* November 1995, Coeur d'Alene, Idaho. Fairfield, Wash.: International Association of Wildland Fire.

Gallup, G., Jr. 1999. *The Gallup poll: Public opinion 1998.* Wilmington, Del.: Scholarly Resources, Inc.

Gartner, F. R., and W. W. Thompson. 1973. Fire in the Black Hills forest-grass ecotone. *Proceedings—Tall Timbers Fire Ecology Conference* 12:37–68. Tall Timbers Research Station, Tallahassee, Florida.

General Accounting Office (GAO). 1999. *Western national forests: A cohesive strategy is needed to address catastrophic wildfire threats.* U.S. General Accounting Office, House of Representatives, Committee on Resources, Report to the Subcommittee on Forests and Forest Health RCED-99-65, Washington, D.C.

Gerritsma, J. 2000. Interviewed by P. Backus. *Missoulian,* 16 October, A1, A7.

Graham, R. T., A. E. Harvey, M. F. Jurgensen, T. B. Jain, J. R. Tonn, and D. S. Page-Dumroese. 1994. *Managing coarse woody debris in forests of the Rocky Mountains.* USDA Forest Service, Intermountain Research Station, Research Paper INT-RP-477, Ogden, Utah.

Gray, A. N., and J. F. Franklin. 1997. Effects of multiple fires on the structure of southwestern Washington forests. *Northwest Science* 71:174–185.

Greenlee, J. M., and J. H. Langenheim. 1990. Historic fire regimes and their relation to vegetation patterns in the Monterey Bay area of California. *American Midland Naturalist* 124:239–253.

Gruell, G. E. 1985a. Indian fires in the Interior West: A widespread influence. Pages 68–74 in J. E. Lotan, B. M. Kilgore, W. C. Fischer, and R. W. Mutch, technical coodinators, *Proceedings: Symposium and workshop on wilderness fire.* USDA Forest Service, Intermountain Forest and Range Experiment Station, General Technical Report INT-182, Ogden, Utah.

———. 1985b. Fire on the early western landscape: An annotated record of wildland fires 1776–1900. *Northwest Science* 59(2):97–107.

———. 2001. *Fire in Sierra Nevada Forests: A photographic interpretation of ecological change since 1849.* Missoula, Mont.: Mountain Press.

Guldin, J. M. 1995. The role of uneven-aged silviculture in the context of ecosystem management. Pages 1–26 in K. O'Hara, ed., *Uneven-aged management: Opportunities, constraints, and methodologies.* Montana Forest and Conservation Experiment Station, University of Montana, Miscellaneous Publication 56, Missoula, Montana.

Habeck, J. R. 1961. The original vegetation of the mid-Willamette Valley, Oregon. *Northwest Science* 35:65–77.

———. 1994. Using General Land Office records to assess forest succession in ponderosa pine/Douglas-fir forests in western Montana. *Northwest Science* 68:69–78.

Haddow, D. 1995. Involvement of wilderness and park fire managers in developing air quality regulations. Pages 80–82 in J. K. Brown, R. W. Mutch, C. W. Spoon, and R. H. Wakimoto, technical coordinators, *Proceedings: Symposium on fire in wilderness and park management.* USDA Forest Service, Intermountain Research Station, General Technical Report INT-GTR-320, Ogden, Utah.

Hardy, C. C., and S. F. Arno, eds. 1996. *The use of fire in forest restoration.* USDA Forest Service, Intermountain Research Station, General Technical Report INT-GTR-341, Ogden, Utah.

Hardy, C. C., R. E. Keane, and C. A. Stewart. 2000. Ecosystem-based management in the lodgepole pine zone. Pages 31–35 in H. Y. Smith, ed., *Proceedings: The Bitterroot Ecosystem Management Research Project: What have we learned symposium.* USDA Forest Service, Rocky Mountain Research Station RMRS-P-17, Ogden, Utah.

Harrington, M. G. 2000. Fire applications in ecosystem management. Pages 21–22 in H. Y. Smith, ed., *Proceedings: The Bitterroot Ecosystem Management Research Project: What have we learned symposium.* USDA Forest Service, Rocky Mountain Research Station RMRS-P-17, Ogden, Utah.

Hartwell, M. G., P. Alaback, and S. F. Arno. 2000. Comparing historic and modern forests on the Bitterroot Front. Pages 11–16 in H. Y. Smith, ed., *Proceedings: The Bitterroot Ecosystem Management Research Project: What have we learned symposium.* USDA Forest Service, Rocky Mountain Research Station, RMRS-P-17, Ogden, Utah.

Heinlein, T. A., W. W. Covington, P. Z. Fule, M. M. Moore, and H. B. Smith. 2000. Development of ecological restoration experiments in fire adapted forests at Grand Canyon National Park. Pages 249–254 in D. N. Cole, S. F. McCool, W. T. Borrie, and J. O'Laughlin, comps., *Proceedings: Wilderness science in a time of change conference.* Vol. 5, *Wilderness ecosystems, threats, and management.* USDA Forest Service, Rocky Mountain Research Station RMRS-P-15-Vol. 5, Ogden, Utah.

Hemstrom, M. A., and J. F. Franklin. 1982. Fire and other disturbances of the forests in Mount Rainier National Park. *Quaternary Research* 18:32–51.

Hermann, R. K., and D. P. Lavender. 1990. *Pseudotsuga menziesii.* Pages 527–540 in R. M. Burns and B. H. Honkala, technical coordinators, *Silvics of North America,* vol. 1: *Conifers.* USDA Forest Service Agriculture Handbook 654. Washington, D.C.: GPO.

Hilbruner, M. 2000. Interviewed by S. Devlin. *Missoulian,* 15 October, A7.

Hirt, P. 2000. Is excellent forestry possible in our national forests? *Distant Thunder* 8:8–9, 12. Santa Fe, N. Mex.: Forest Stewards Guild.

Houston, D. B. 1982. *The northern Yellowstone elk: Ecology and management.* New York: Macmillan.

Hoxie, G. L. 1910. How fire helps forestry. *Sunset* 34:145–151.

Hull, J. D. 1987. A just war. *Time,* 26 October, 23–24, 29.

Hungerford, R. D., M. G. Harrington, W. H. Frandsen, K. C. Ryan, and G. J. Niehoff. 1991. Influence of fire on factors that affect site productivity. Pages 32–50 in A. E. Harvey and L. F. Neuenschwander, comps., *Proceedings: Management and productivity of western-montane forest soils.* USDA Forest Service Intermountain Research Station, General Technical Report INT-280, Ogden, Utah.

Hutcherson, A. 1991. Tree farmers like being tree farmers. *Tree Farmer,* summer, 14–15.

Illg, C., and G. Illg. 1994. The ponderosa and the flammulated. *American Forests* 100(3/4):36–37, 58.

———. 1997. Forest fire, water quality, and the incident at Buffalo Creek. *American Forests* 103(1):33–35.

Intermountain Fire Council. 1985. *Prescribed fire by aerial ignition.* Proceedings of a workshop, Missoula, Montana, Oct. 30–Nov. 1, 1984. Montana Department of Natural Resources and Conservation, Forestry Division.

Jay, J., M. Arno, and N. Arno. 1998. Is carefully applied silviculture worth more than lowest-cost logging? *Montana Tree Farm News* 14:6–7. (Montana Tree Farm Program, P.O. Box 17276, Missoula, MT, 59808.)

Johannessen, C. L. 1971. The vegetation of the Willamette Valley. *Annals of the Association of American Geographers* 61(2):286–302.

Jones, J. G., and J. D. Chew. 1999. Applying simulation and optimization to evaluate the effectiveness of fuel treatments for different fuel conditions at landscape scales. Pages 89–96 in L. F. Neuenschwander, K. C. Ryan, and G. E. Gollberg, eds., *Proceedings from the Joint Fire Sciences Conference and Workshop,* Boise, Idaho, June 12–17, 1999. Moscow: International Association of Wildland Fire and University of Idaho.

Juvan, G., and B. Habeck. 2001. Prescribed burns benefit the forest. *Independent Record (Helena, Montana)*, 5 April, 3c–3e.

Kalabokidis, K. D., and R. H. Wakimoto. 1992. Prescribed burning in uneven-aged stand management of ponderosa pine/Douglas-fir forests. *Journal of Environmental Management* 34:221–235.

Kauffman, J. B. 1990. Ecological relationships of vegetation and fire in Pacific Northwest forests. Pages 39–52 in J. D. Walstad, S. R. Radosevich, and D. V. Sandberg, *Natural and prescribed fire in Pacific Northwest forests.* Corvallis: Oregon State University Press.

Kay, C. E. 1993. Aspen seedlings in recently burned areas of Grand Teton and Yellowstone National Parks. *Northwest Science* 67:94–104.

Keane, R. E., S. F. Arno, and J. K. Brown. 1990. Simulating cumulative fire effects in ponderosa pine/Douglas-fir forests. *Ecology* 71:189–203.

Keane, R. E., S. F. Arno, and C. A. Stewart. 2000. Ecosystem-based management in the whitebark pine zone. Pages 36–40 in Y. H. Smith, ed., *Proceedings: The Bitterroot Ecosystem Management Research Project: What have we learned symposium.* USDA Forest Service, Rocky Mountain Research Station RMRS-P-17, Ogden, Utah.

Keane, R. E., K. Ryan, and S.W. Running. 1996. Simulating the effect of fire on northern Rocky Mountain landscapes using the ecological process model Fire-BGC. *Tree Physiology* 16:319–331.

Keifer, M., N. L. Stephenson, and J. Manley. 2000. Prescribed fire as the minimum tool for wilderness forest and fire regime restoration: A case study from the Sierra Nevada, California. Pages 266–269 in D. N. Cole, S. F. McCool, W. T. Borrie, and J. O'Laughlin, comps., *Wilderness science in a time of change conference.* Vol. 5, *Wilderness ecosystems, threats, and management.* USDA Forest Service, Rocky Mountain Research Station, Proceedings RMRS-P-15-Vol. 5, Ogden, Utah.

Kent, B., W. D. Shepperd, and D. J. Shields. 2000. The Colorado Front Range Ecosystem Management Research Project: Accomplishments to date. Pages 119–124 in H. Y. Smith, ed., *The Bitterroot Ecosystem Management Research Project: What we have learned symposium.* USDA Forest Service, Rocky Mountain Research Station, Proceedings RMRS-P-17, Ogden, Utah.

Kiester, E., Jr. 1999. Town buries the axe. *Smithsonian* 30(4):70–79.

Kilgore, B. M. 1973. The ecological role of fire in Sierran conifer forests: Its application to national park management. *Quaternary Research* 3(3):496–513.

———. 1976. *From fire control to fire management: An ecological basis for policies.* Transactions of the 41st North American Wildlife and Natural Resources Conference. Washington, D.C.: Wildlife Management Institute.

———. 1986. The role of fire in wilderness: A state-of-knowledge review. Pages 70–103 in R. C. Lucas, ed., *Proceedings: National wilderness research conference: Issues, state-of-knowledge, future directions.* USDA Forest Service, Intermountain Research Station, General Technical Report INT-220, Ogden, Utah.

Kilgore, B. M., and G. S. Briggs. 1972. Restoring fire to high elevation forests in California. *Journal of Forestry* 70(5):266–271.

Kilgore, B. M., and G. A. Curtis. 1987. Guide to understory burning in ponderosa pine–larch–fir forests in the Intermountain West. USDA Forest Service, Intermountain Research Station, General Technical Report INT-233, Ogden, Utah.

Knight, D. H. 1987. Parasites, lightning, and the vegetation mosaic in wilderness landscapes. Pages 59–82 in M. G. Turner, ed., *Landscape heterogeneity and disturbance.* New York: Springer-Verlag.

———. 1991. The Yellowstone fire controversy. Pages 87–103 in R. B. Keiter and M. S. Boyce, eds., *The greater Yellowstone ecosystem: Redefining America's wilderness heritage.* New Haven, Conn.: Yale University Press.

———. 1996. The ecological implications of fire in Greater Yellowstone: A summary. Pages 233–235 in J. M. Greenlee, ed., *The ecological implications of fire in Greater Yellowstone.* Proceedings: Second Biennial Conference on the Greater Yellowstone Ecosystem. Fairfield, Wash.: International Association of Wildland Fire.

Koch, E. 1935. The passing of the Lolo Trail. *Journal of Forestry* 33(2):98–104.

———. 1942. *History of the 1910 forest fires in Idaho and western Montana.* USDA Forest Service, Northern Region, Mimeographed report, Missoula, Montana.

———. 1998. *Forty years a forester: 1903–1943.* Missoula, Mont.: Mountain Press.

Küchler, A. W. 1964. Potential natural vegetation of the conterminous United States. Map, scale 1:3,168,000. American Geographical Society, New York City.

Kurth, L. L. 1996. Examples of fire restoration in Glacier National Park. Pages 54–55 in C. C. Hardy, and S. F. Arno, eds., *The use of fire in forest restoration.* USDA Forest Service, Intermountain Research Station, General Technical Report INT-341, Ogden, Utah.

Lamb, S. H. 1943. [Letter in response to:] Fire as an ecological and silvicultural factor. *Journal of Forestry* 41: 294–295.

LeBarron, R. K. 1957. Silvicultural possibilities of fire in northeastern Washington. *Journal of Forestry* 55:627–630.

Leiberg, J. B. 1899. *Bitterroot Forest Reserve.* U.S. Geological Survey 19th Annual Report, Part 5:253–282.

Leopold, A. 1924. Grass, brush, timber, and fire in southern Arizona. *Journal of Forestry* 22(6):1–10.

Leopold, A. S., S. A. Cain, C. Cottam, I. N. Gabrielson, and T. L. Kimball. 1963. *Wildlife management in the national parks.* Transactions of the North American Wildlife and Natural Resources Conference 28:28–45.

Lewis, H. T. 1973. *Patterns of Indian burning in California: Ecology and ethnohistory.* Anthropology Paper No.1. Ramona, Calif.: Ballena Press.

———. 1985. Why Indians burned: Specific versus general reasons. Pages 75–80 in J. E. Lotan, B. M. Kilgore, W. C. Fischer, and R. W. Mutch, technical coordinator, *Proceedings: Symposium and workshop on wilderness fire.* USDA Forest Service, Intermountain Forest and Range Experiment Station, General Technical Report INT-182, Ogden, Utah.

Little, E. L., Jr. 1971. *Atlas of United States trees,* vol. 1: *Conifers and important hardwoods.* USDA Forest Service, Miscellaneous Publication no. 1146. Washington, D.C.: GPO.

Little, J. B. 1998. The woods: Reclaiming the neighborhood. *American Forests* 103(4):12–13, 39–41.

————. 2000. Stewardships trial by forests. *American Forests* 106(3):49–55.

Loveridge, E. W. 1935. The opposite point of view. *Journal of Forestry* 33(2):105–110.

Lutts, R. H. 1992. The trouble with Bambi: Walt Disney's Bambi and the American vision of nature. *Forest and Conservation History* 36:160–171.

Lyon, L. J., M. H. Huff, R. G. Hooper, E. S. Telfer, D. S. Schreiner, and J. K. Smith. 2000. *Wildland fire in ecosystems: Effects of fire on fauna.* USDA Forest Service, Rocky Mountain Research Station, General Technical Report RMRS-GTR-42-Vol. 1, Ogden, Utah.

Mahoney, D. L., and R. E. Gresswell. 1998. Short-term postfire variation of physical habitat, fish populations, and associated aquatic communities in large streams of Yellowstone National Park. *Yellowstone Science* 6(2):39.

Manning, R. 2001. Friendly fire. *Sierra* 86(1):40–41, 110.

Martin, R. E. 1990. Goals, methods, and elements of prescribed burning. Pages 55–66 in J. D. Walstad, S. R. Radosevich, and D. V. Sandberg, eds., *Natural and prescribed fire in Pacific Northwest forests.* Corvallis: Oregon State University Press.

Martin, R. E., J. B. Kauffman, and J. D. Landsberg. 1989. Use of prescribed fire to reduce wildfire potential. Pages 17–22 in N. H. Berg, technical coordinator, *Proceedings: Symposium on fire and watershed management.* USDA Forest Service, Pacific Southwest Forest and Range Experiment Station, General Technical Report PSW-109, Berkeley, California.

Mattson, D. J., K. C. Kendall, and D. P. Reinhart. 2001. Whitebark pine, grizzly bears, and red squirrels. Pages 121–136 in D. F. Tomback, S. F. Arno, and R. E. Keane, eds., *Whitebark pine communities: Ecology and restoration.* Washington, D.C.: Island Press.

McBride, J. R. 1983. Analysis of tree rings and fire scars to establish fire history. *Tree-Ring Bulletin* 43:51–67.

McClelland, B. R., S. S. Frissell, W. C. Fischer, and C. H. Halvorson. 1979. Habitat management for hole-nesting birds in forests of western larch and Douglas-fir. *Journal of Forestry* 77(8):480–483.

McCool, S. F., and G. H. Stankey. 1986. *Visitor attitudes toward wilderness fire management policy 1971–84.* USDA Forest Service, Intermountain Research Station, Research Paper INT-357, Ogden, Utah.

McCune, B. 1983. Fire frequency reduced two orders of magnitude in the Bitterroot Canyons, Montana. *Canadian Journal of Forestry Research* 13:212–218.

McDonald, P. M. 1990. *Quercus kelloggii Newb.* California black oak. Pages 661–671 in R. M. Burns and B. H. Honkala, technical coordinators, *Silvics of North America*, vol. 2: *Hardwoods* USDA Forest Service, Agriculture Handbook 654: GPO.

McKee, R. 2000. Personal communication with author, December.

McKelvey, K. S., C. N. Skinner, C. Chang, D. C. Erman, S. J. Husari, D. J. Parsons, J. W. van Wagtendonk, and C. P. Weatherspoon. 1996. An overview of fire in the Sierra Nevada. Pages 1033–1040 in *Sierra Nevada Ecosystem Project: Final report to Congress*, vol. 2. Davis: Centers for Water and Wildland Resources, University of California.

McPhee, J. 1988. The control of nature: Los Angeles against the mountains. *New Yorker*, 26 September, 45–78, and 3 October, 72–90.

Means, J. E., J. H. Cissel, and F. J. Swanson. 1996. *Fire history and landscape restoration in Douglas-fir ecosystems of western Oregon*. Pages 61–67 in C. C. Hardy and S. F. Arno, eds., *The use of fire in forest restoration*. USDA Forest Service, Intermountain Research Station, General Technical Report INT-341, Ogden, Utah.

Means, J. E., T. Spies, S. Chen, J. Kertis, and P. Teensma. 1996. Forests of the Oregon Coast Range: Considerations for ecological restoration. Pages 68–71 in C. C. Hardy and S. F. Arno, eds., *The use of fire in forest restoration*. USDA Forest Service, Intermountain Research Station, General Technical Report INT-341, Ogden, Utah.

Meeks, P. 2001. Teamwork heats up. *American Forests* 107(1):7–9.

Miles, S. R., D. M. Haskins, and D. W. Ranken. 1989. Emergency burn rehabilitation: Cost, risk, and effectiveness. Pages 97–102 in N. H. Berg, technical coordinator, *Proceedings: Symposium on fire and watershed management*. USDA Forest Service, Pacific Southwest Forest and Range Experiment Station, General Technical Report PSW-109, Berkeley, California.

Miller, C. 2000. The pivotal decade: American forestry in the 1870s. *Journal of Forestry* 98(11):6–10.

Miller, C., and D. L. Urban. 2000. Modeling the effects of fire management alternatives on Sierra Nevada mixed-conifer forests. *Ecological Applications* 10(1):85–94.

Miller, M. 2000. Fire autecology. Pages 9–34 in J. K. Brown, ed., *Effects of fire on flora*. USDA Forest Service, Rocky Mountain Research Station, General Technical Report RM-42-Vol. 2, Fort Collins, Colorado.

Millspaugh, S. H., C. Whitlock, and P. J. Bartlein. 2000. Variations in fire frequency and climate over the past 17,000 yr in central Yellowstone National Park. *Geology* 28(3):211–214.

Minore, D. 1979. *Comparative autecological characteristics of northwestern tree species: A literature review*. USDA Forest Service, General Technical Report PNW-87, Portland, Oregon.

Minshall, G. W., and J. T. Brock. 1991. Pages 123–136 in R. B. Keiter and M. S. Boyce, eds., *The Greater Yellowstone Ecosystem: Redefining America's wilderness heritage*. New Haven, Conn.: Yale University Press.

Minshall, G. W., C. T. Robinson, and T. V. Royer. 1998. Stream ecosystem response to the 1988 Yellowstone wildfires: A ten-year review. *Yellowstone Science* 6(2):41.

Monnig, E., and J. Byler. 1992. *Forest health and ecological integrity in the northern Rockies.* USDA Forest Service, Northern Region, FPM Report 92-7, Missoula, Montana.

Moore, B. 1996. *The Lochsa story: Land ethics in the Bitterroot Mountains.* Missoula, Mont.: Mountain Press.

Moore, M. M., W. W. Covington, and P. Z. Fulé. 1999. Reference conditions and ecological restoration: A southwestern ponderosa pine perspective. *Ecological Applications* 9(4):1266–1277.

Moore, P. 1999. Pacific spirit—the forest reborn. *Journal of Sustainable Forestry* 9(1/2):135–155.

Morrison, P. H., and F. J. Swanson. 1990. *Fire history and pattern in a Cascade Range landscape.* USDA Forest Service, Pacific Northwest Forest and Range Experiment Station, General Technical Report PNW-254, Portland, Oregon.

Mutch, R. W. 2001. Practice, poetry, and policy: Will we be better prepared for the fires of 2006? *Bugle* 18(2):61–64.

Mutch, R. W., S. F. Arno, J. K. Brown, C. E. Carlson, R. D. Ottmar, and J. L. Peterson. 1993. *Forest health in the Blue Mountains: A management strategy for fire-adapted ecosystems.* USDA Forest Service, Pacific Northwest Research Station, General Technical Report PNW-310, Portland, Oregon.

Native Forest Network. 2001. *Wildfire: A natural part of life in the American West.* Missoula, Mont.: Native Forest Network.

Nelson, T. C. 1979. Fire management policy in the national forests: A new era. *Journal of Forestry* 77:723–725.

New York Times. 1889. Imprisoned by fire. 14 August, 1.

———. 1889. Forest and prairie fires. 15 August, 5.

———. 1889. The Montana forest fires. 20 August, 3.

Norton, H. H. 1979. The association between anthropogenic prairies and important food plants in western Washington. *Northwest Anthropological Research Notes* 13:175–200.

O'Laughlin, J., J. G. MacCracken, D. L. Adams, S. C. Bunting, K. A. Blatner, and C. E. Keegan III. 1993. *Forest health conditions in Idaho: Executive summary.* Idaho Forest, Wildlife and Range Policy Analysis Group, Report No. 11. Moscow: University of Idaho.

Oliver, W. W. 2001. *Ecological research at the Black Mountain Experimental Forest in northeastern California.* USDA Forest Service, Pacific Southwest Research Station, General Technical Report 179, Berkeley, California.

Olson, D. L. 2000. Fire in riparian zones: A comparison of historical fire occurrence in riparian and upslope forests in the Blue Mountains and southern Cascades of Oregon. M.S. thesis, University of Washington, Seattle.

Omi, P. N. 1996. Landscape-level fuel manipulations in Greater Yellowstone: Opportunities and challenges. Pages 7–14 in J. M. Greenlee, ed., *The ecological implications of fire in Greater Yellowstone.* Proceedings: Second Biennial

Conference on the Greater Yellowstone Ecosystem. Fairfield, Wash.: International Association of Wildland Fire.

Omi, P. N., and K. D. Kalabokidis. 1991. Fire damage on extensively vs. intensively managed forest stands within the North Fork Fire, 1988. *Northwest Science* 65(4):149–157.

Ottmar, R. D., M. D. Schaaf, and E. Alvarado. 1996. Smoke considerations for using fire in maintaining healthy forest ecosystems. Pages 24–28 in C. C. Hardy and S. F. Arno, eds., *The use of fire in forest restoration.* USDA Forest Service, Intermountain Research Station, General Technical Repot INT-GTR-341, Ogden, Utah.

Parsons, D. J. 2000. The challenge of restoring natural fire to wilderness. Pages 276–282 in D. N. Cole, S. F. McCool, W. T. Borrie, and J. O'Loughlin, comps., *Proceedings: Wilderness science in a time of change conference.* Vol. 5, *Wilderness ecosystems, threats, and management.* USDA Forest Service, Rocky Mountain Research Station. RMRS-P-Vol. 5, Ogden, Utah.

Parsons, D. J., and P. B. Landres. 1998. Restoring natural fire to wilderness: How are we doing? *Proceedings—Tall Timbers Fire Ecology Conference* 20:366–373. Tall Timbers Research Station, Tallahassee, Florida.

Perkins, S. 2001. A nation aflame: After last year's conflagrations, can we learn to live with wildfire? *Science News* 159:120–121.

Perlin, J. 1991. *A forest journey: The role of wood in the development of civilization.* Cambridge, Mass.: Harvard University Press.

Peterson, D. L., and R. D. Hammer. 2001. From open to closed canopy: A century of change in a Douglas-fir forest, Orcas Island, Washington. *Northwest Science* 75:262–269.

Pinchot, G. 1899. The relation of forests and forest fires. *National Geographic Magazine* 10:393–403.

———. [1947] 1972. *Breaking new ground.* Reprint, Seattle: University of Washington Press.

Pollet, J., and P. N. Omi. 1999. *Effect of thinning and prescribed burning on wildfire severity in ponderosa pine forests.* USDA Forest Service, Rocky Mountain Research Station, Fire Sciences Laboratory, Contract completion report: Agreement INT-95075-RJVA, Missoula, Montana.

Powell, D. S., J. L. Faulkner, D. R. Darr, Z. Zhu, and D. W. MacCleery. 1993. *Forest resources of the United States, 1992.* USDA Forest Service, Rocky Mountain Forest and Range Experiment Station, General Technical Report RM-234, Fort Collins, Colorado.

Powell, J. W. 1891. *Testimony to Congress: Eleventh annual report of the U.S. Geological Survey, 1889–1890,* part 2: *Irrigation,* 207–208. Washington, D.C.: GPO.

Pyne, S. J. 1982. *Fire in America: A cultural history of wildland and rural fire.* Princeton: Princeton University Press.

———. 1984. *Introduction to wildland fire.* New York: John Wiley and Sons.

———. 1997. *World fire: The culture of fire on Earth.* Seattle: University of Washington Press.

————. 2001a. *Year of the fires: The story of the great fires of 1910.* New York: Viking Penguin.

————. 2001b. The perils of prescribed fire: A reconsideration. *Natural Resources Journal* 41:1–8.

Quigley, T. M., R. W. Haynes, and R. T. Graham, technical eds. 1996. *Integrated scientific assessment for ecosystem management in the interior Columbia Basin.* USDA Forest Service, Pacific Northwest Research Station, General Technical Report PNW-382, Portland, Oregon.

Reed, M. J. 1974. Ceanothus. Pages 284–290 in C. S. Schopmeyer, technical coordinator, *Seeds of woody plants in the United States.* USDA Forest Service, Agriculture Handbook 450. Washington, D.C.: GPO.

Reid, B. 1998. A clearing in the forest: New approaches to managing America's woodlands. *Nature Conservancy* 48(6):18–24.

Reider, D. A. 1988. California conflagration. *Journal of Forestry* 86(1):5–8.

Reynolds, R. T., W. M. Block, and D. A. Boyce, Jr. 1996. Using ecological relationships of wildlife as templates for restoring southwestern forests. Pages 34–42 in W. Covington and P. K. Wagner, technical coordinators, *Conference on adaptive ecosystem restoration and management.* USDA Forest Service, Rocky Mountain Forest and Range Experiment Station, General Technical Report RM-278, Fort Collins, Colorado.

Rice, P. 2000. *Restoration of native plant communities infested by invasive weeds: Sawmill RNA.* Page 134 in H. Y. Smith, ed., *Proceedings: The Bitterroot Ecosystem Management Research Project: What have we learned symposium.* USDA Forest Service, Rocky Mountain Research Station, Proceedings RMRS-P-17, Ogden, Utah.

Robock, A. 1988. Enhancement of surface cooling due to forest fire smoke. *Science* 242:911–913.

Romme, W. H. 1982. Fire and landscape diversity in subalpine forests of Yellowstone National Park. *Ecological Monographs* 52:199–221.

Romme, W. H., and D. Despain. 1989. Historical perspective on the Yellowstone fires of 1988. *Bioscience* 39:695–699.

Rothermel, R. C., R. A. Hartford, and C. H. Chase. 1994. *Fire growth maps for the 1988 Greater Yellowstone Area fires.* USDA Forest Service, Intermountain Research Station, General Technical Report INT-304, Ogden, Utah.

Roy, D. F. 1980. Redwood. Pages 109–110 in F. H. Eyre, ed., *Forest cover types of the United States and Canada.* Washington, D.C.: Society of American Foresters.

Ryan, K. C., and N. V. Noste. 1985. *Evaluating prescribed fires.* Pages 230–238 in J. E. Lotan, B. M. Kilgore, W. C. Fischer, and R. W. Mutch, technical coordinators, *Proceedings—Symposium and workshop on wilderness fire.* USDA Forest Service, Intermountain Forest and Range Experiment Station, General Technical Report INT-182, Ogden, Utah.

Sackett, S. S., S. M. Haase, and M. G. Harrington. 1996. Lessons learned from fire use for restoring Southwestern ponderosa pine ecosystems. Pages 54–61 in W. Covington and P. K. Wagner, technical coordinators, *Conference on adap-*

tive ecosystem restoration and management. USDA Forest Service, Rocky Mountain Forest and Range Experiment Station, General Technical Report RM-278, Fort Collins, Colorado.

Salwasser, H., and R. D. Pfister. 1994. Ecosystem management: From theory to practice. Pages 150–161 in W. W. Covington, and L. F. DeBano, technical coordinators, *Sustainable ecological systems: Implementing an ecological approach to land management.* USDA Forest Service, Rocky Mountain Forest and Range Experiment Station, General Technical Report RM-247, Fort Collins, Colorado.

Schiff, A. 1962. *Fire and water: Scientific heresy in the Forest Service.* Cambridge, Mass.: Harvard University Press.

Schmidt, W. C., and R. H. Wakimoto. 1988. *Cultural practices that can reduce fire hazards to homes in the interior West.* Pages 131–141 in W. C. Fischer and S. F. Arno, comps., *Protecting people and homes from wildfire in the Interior West: Proceedings of the symposium and workshop.* USDA Forest Service, Intermountain Research Station, General Technical Report INT-251, Ogden, Utah.

Schoennagel, T. L., and D. M. Waller. 1999. Understory responses to fire and artificial seeding in an eastern Cascades *Abies grandis* forest, U.S.A. *Canadian Journal of Forestry Research* 29:1393–1401.

Scott, J. H. 1998. *Fuel reduction in residential and scenic forests: A comparison of three treatments in a western Montana ponderosa pine stand.* USDA Forest Service, Rocky Mountain Research Station, Research Paper RMRS-5, Ogden, Utah.

Shinneman, D. J., and W. L. Baker. 1997. Nonequilibrium dynamics between catastrophic disturbances and old-growth forests in ponderosa pine landscapes of the Black Hills. *Conservation Biology* 11:1276–1288.

Shoemaker, T. 1922. Twenty years of protection in forests. *Daily Missoulian,* Souvenir Edition, 20 July, 7–9. Missoula, Montana.

Show, S. B., and E. I. Kotok. 1924. *The role of fire in the California pine forests.* USDA Bulletin No. 1294. Washington, D.C.: GPO.

Sierra Club. 2001. *Forest fires: Beyond the heat and hype.* Washington, D.C.: Sierra Club.

Simmerman, D. G., and W. C. Fischer. 1990. Prescribed fire practices in the Interior West: A survey. Pages 66–75 in M. E. Alexander and G. F. Bisgrove, technical coordinators, *The art and science of fire management.* Proceedings of Interior West Fire Council, October 24–27, 1988. Forestry Canada, Northern Forestry Center, Edmonton, Alberta.

Singer, F. J., and P. Schullery. 1989. Yellowstone wildlife: Populations in process. *Western Wildlands* 15(2):18–22.

Sirucek, D. 1988. The North Hills Fire-Erosion Event. *Forestry Abstracts* 49(12):8578.

Skinner, C. N. 2001. Personal communication. Written commentary to author, March.

Skinner, C. N., and C. Chang. 1996. Fire regimes, past and present. Pages 1041–1069 in *Sierra Nevada Ecosystem Project: Final report to Congress,* vol. 2.

Davis: Centers for Water and Wildland Resources, University of California.

Smith, C. 1992. *Media and apocalypse: News coverage of the Yellowstone forest fires, Exxon Valdez oil spill, and Loma Prieta earthquake.* Westport, Conn.: Greenwood Press.

———. 1995. Fire issues and communication by the media. Pages 65–69 in J. K. Brown, R. W. Mutch, C. W. Spoon, and R. H. Wakimoto, technical coordinators, *Proceedings: Symposium on fire in wilderness and park management.* USDA Forest Service, Intermountain Research Station, General Technical Report INT-GTR-320, Ogden, Utah.

Smith, H. Y. 2000a. Factors affecting ponderosa pine snag longevity. Pages 223–229 in *Proceedings of the Society of American Foresters 1999 National Convention.* Bethesda, Md.: Society of American Foresters.

———. 2000b. *Wildlife habitat considerations.* Pages 26–27 in H. Y. Smith, ed., *Proceedings—The Bitterroot Ecosystem Management Research Project: What we have learned.* USDA Forest Service, Rocky Mountain Research Station RMRS-P-17:26–27, Ogden, Utah.

———, ed. 2000c. *Proceedings—The Bitterroot Ecosystem Management Research Project: What we have learned.* USDA Forest Service, Rocky Mountain Research Station RMRS-P-17, Ogden, Utah.

Smith, J. K., and W. C. Fischer. 1997. *Fire ecology of the forest habitat types of northern Idaho.* USDA Forest Service, Intermountain Research Station, General Technical Report INT-363, Ogden, Utah.

Starr, L., J. McIver, and T. M. Quigley. 2000. Sustaining the land, people, and economy of the Blue Mountains: The Blue Mountains Natural Resources Institute. Pages 125–128 in H. Y. Smith, ed., *Proceedings—The Bitterroot Ecosystem Management Research Project: What we have learned.* USDA Forest Service, Rocky Mountain Research Station, RMRS-P-17, Ogden, Utah.

Stephens, S. L. 1998. Evaluation of the effects of silviculture and fuels treatments on potential fire behavior in Sierra Nevada mixed-conifer forests. *Forest Ecology and Management* 105:21–35.

Stephenson, N. L. 1999. Reference conditions for giant sequoia forest restoration: Structure, process, and precision. *Ecological Applications* 9(4):1253–1265.

Stewart, F. J. 1998. Transforming how the USDA Forest Service markets timber sales. Pages 146–151 in *Forest management into the next century: What will make it work?* Madison, Wis.: Forest Products Society.

Stickney, P. F. 1990. Early development of vegetation following holocaustic fire in Northern Rocky Mountain forests. *Northwest Science* 64(5):243–246.

Stokes, M.A., and T. L. Smiley. 1968. *An introduction to tree-ring dating.* Chicago: University of Chicago Press.

Stuart, J. D. 1987. Fire history of an old-growth forest of *Sequoia sempervirens* (Taxodiaceae) forest in Humboldt Redwoods State Park, California. *Madrono* 34:128–141.

Sudworth, G. B. 1908. *Forest trees of the Pacific Slope.* USDA Forest Service. Washington, D.C.: GPO.

Sutherland, E. K. 1997. History of fire in a southern Ohio second-growth mixed-oak forest. Pages 172–183 in S. G. Pallardy, R. A. Cecich, H. E. Garrett, and P. S. Johnson, eds., *Proceedings: 11th Central Hardwood Forest Conference.* USDA Forest Service, North Central Forest Experiment Station, General Technical Report NC-188, St. Paul, Minnesota.

Swetnam, T. W. 1984. Peeled ponderosa pine trees: A record of inner bark utilization by Native Americans. *Journal of Ethnology* 4(2):177–190.

———. 1993. Fire history and climate change in giant sequoia groves. *Science* 262:885–889.

Swetnam, T. W., and C. H. Baisan. 1996. Historical fire regime patterns in the southwestern United States since A.D. 1700. Pages 11–32 in C. D. Allen, ed., *Proceedings of the 2nd La Mesa Fire Symposium.* USDA Forest Service, Rocky Mountain Forest and Range Experiment Station, General Technical Report RM-GTR-286, Fort Collins, Colorado.

Swetnam, T. W., C. D. Allen, and J. L. Betancourt. 1999. Applied historical ecology: Using the past to manage for the future. *Ecological Applications* 9(4):1189–1206.

Sydoriak, C. A., C. D. Allen, and B. F. Jacobs. 2000. Would ecological landscape restoration make the Bandelier Wilderness more or less of a Wilderness? Pages 209–215 in D. N. Cole, S. F. McCool, W. T. Borrie, and J. O'Loughlin, comps., *Wilderness science in a time of change conference.* Vol. 5, *Wilderness ecosystems, threats, and management.* USDA Forest Service, Rocky Mountain Research Station, RMRS-P-15-Vol. 5, Ogden, Utah.

Tande, G. F. 1979. Fire history and vegetation patterns of coniferous forests in Jasper National Park, Alberta. *Canadian Journal of Botany* 57:1912–1931.

Taylor, A. H., and C. N. Skinner. 1998. Fire history and landscape dynamics in a late-successional reserve, Klamath Mountains, California, USA. *Forest Ecology and Management* 111:285–301.

Taylor, O. J. 1989. *Montana 1889: The centennial news melange.* Virginia City, Mont.: O. J. Taylor, P.O. Box 97, Virginia City, MT 59755. [Privately published.]

Tewksbury, J. J., T. E. Martin, S. J. Hejl, T. S. Redman, and F. J. Wheeler. 1999. Cowbirds in a western valley: Effects of landscape structure, vegetation, and host density. *Studies in Avian Biology* 18:23–33.

Tinner, W., P. Hubschmid, M. Wehrli, B. Ammann, and M. Conedera. 1999. Long-term forest fire ecology and dynamics in southern Switzerland. *Journal of Ecology* 87:273–289.

Tomback, D. F. 2001. Clark's nutcracker: Agent of regeneration. Pages 89–104 in D. F. Tomback, S. F. Arno, and R. E. Keane, eds., *Whitebark pine communities ecology and restoration.* Washington, D.C.: Island Press.

Tomback, D. F., S. F. Arno, and R. E. Keane. 2001. The compelling case for management intervention. Pages 3–25 in D. F. Tomback, S. F. Arno, and R. E. Keane, eds., *Whitebark pine communities ecology and restoration.* Washington, D.C.: Island Press.

Trewartha, G. T. 1968. *An introduction to climate.* 4th ed. New York: McGraw-Hill.

Turner, M. G., W. H. Romme, and R. H. Gardner. 1999. Prefire heterogeneity, fire severity, and early postfire plant reestablishment in subalpine forests of Yellowstone National Park, Wyoming. *International Journal of Wildland Fire* 9(1):21–36.

Turner, N. J. 1999. Time to burn: Traditional use of fire to enhance resource production by Aboriginal peoples in British Columbia. Pages 185–218 in R. Boyd, ed., *Indians, fire, and the land in the Pacific Northwest.* Corvallis: Oregon State University Press.

Tveten, R. K., and R. W. Fonda. 1999. Fire effects on prairies and oak woodlands on Fort Lewis, Washington. *Northwest Science* 73:145–158.

USDA Forest Service. 2001. National fire plan, budget summary. Online: www.na.fs.fed.us/nfp/overview. 1 August.

USDA-USDI. 1989. Final report and recommendations of the Fire Management Policy Review Team and summary of public comments; notice. *Federal Register* 54 (115):25660–25678, 16 June.

Uuskoski, T. 2001. Fiberpac bundles bioenergy for power plants. *Timberjack News* 1:12–13.

Vanhorn, F. 2001. Personal communication. Fire management officer, Glacier National Park, West Glacier, Montana.

Wakimoto, R. H. 1989. National fire management policy: A look at the need for change. *Western Wildlands* 15(2):35–39.

Washington Department of Ecology. 1997. *Natural event action plan for wildfire particulate matter in Chelan County, Washington.* Olympia: Washington Department of Ecology.

Washington State Department of Natural Resources. 1998. *Our changing nature: Natural resource trends in Washington State.* Olympia: Washington State Department of Natural Resources.

Weatherspoon, C. P. 1990. Giant sequoia. Pages 552–562 in R. M. Burns and B. H. Honkala, technical coordinators, *Silvics of North America,* vol. 1: *Conifers.* USDA Forest Service, Agriculture Handbook 654. Washington, D.C.: GPO.

Weatherspoon, C. P., and C. N. Skinner. 1995. An assessment of factors associated with damage to tree crowns from the 1987 wildfires in northern California. *Forest Science* 41(3):430–451.

———. 1996. Landscape-level strategies for forest fuel management. Pages 1471–1492 in *Sierra Nevada Ecosystem Project: Final Report to Congress,* vol. 2. Davis: University of California, Centers for Water and Wildland Resources.

Weaver, H. 1943. Fire as an ecological and silvicultural factor in the ponderosa pine region of the Pacific Slope. *Journal of Forestry* 41(1):7–14.

———. 1968. Fire and its relationship to ponderosa pine. *Proceedings—Tall Timbers Fire Ecology Conference* 7:127–149. Tall Timbers Research Station, Tallahassee, Florida.

Webb, D. 2000. Meeting notes. Interagency Wildfire Management Team. May

24, 2000. [From R. G. Balice, Ecology Group, Los Alamos National Laboratory, New Mexico.]

Wellner, C. A. 1970. Fire history in the northern Rocky Mountains. Pages 42–64 in *Proceedings: The role of fire in the Intermountain West.* Missoula: University of Montana, School of Forestry.

Wheeler, F. J., T. S. Redman, and J. J. Tewksbury. 1997. Montana riparian habitat for birds: Islands of diversity, retreats for survival. Blue Mountains Natural Resource Institute, LaGrande, Oregon. *Natural Resource News* 7(2):10–11.

White, A. S. 1985. Presettlement regeneration patterns in a southwestern ponderosa pine stand. *Ecology* 66:589–594.

Wickman, B. E. 1992. *Forest health in the Blue Mountains: The influence of insects and diseases.* USDA Forest Service, Pacific Northwest Research Station, General Technical Report PNW-295, Portland, Oregon.

Wilkinson, T. 2001. Prometheus unbound. *Nature Conservancy* 51(3):12–20.

Williams, J. 1995. Firefighter safety in changing forest ecosystems. *Fire Management Notes* 55(3):6–8.

———. 2001. New fire manager takes hard lessons to D.C. Interview by Sherry Devlin, *Missoulian,* 17 June, A1, A5.

Wills, R. D., and J. D. Stuart. 1994. Fire history and stand development of a Douglas-fir/hardwood forest in northern California. *Northwest Science* 68:205–212.

Wright, H. A., and A. W. Bailey. 1982. *Fire ecology: United States and southern Canada.* New York: John Wiley and Sons.

Wuerthner, G. 2000. Big blazes needed. *Forest Magazine,* 12 September. Online: www.afseee.org/fsnews. Forest Service Employees for Environmental Ethics, Eugene, Oregon. 1 May.

Zack, S., W. L. Laudenslayer, T. L. George, C. Skinner, and W. Oliver. 1999. A prospectus on restoring late successional forest structure to eastside pine ecosystems through large-scale, interdisciplinary research. In J. E. Cook and B. P. Oswald, comps., *Proceedings: First Biennial North American Forest Ecology Workshop,* June 24–26, 1997, North Carolina State University, Raleigh, North Carolina. Bethesda, Md.: Society of American Foresters.

Zimmerman, G. T., and D. L. Bunnell. 2000. The federal wildland fire policy: Opportunities for wilderness fire management. Pages 288–297 in D. N. Cole, S. F. McCool, W. T. Borrie, and J. O'Loughlin, comps., *Wilderness science in a time of change conference.* Vol. 5, *Wilderness ecosystems, threats, and management.* USDA Forest Service, Rocky Mountain Research Station, RMRS-P-15-Vol. 5, Ogden, Utah.

Zimmerman, G. T., and P. N. Omi. 1998. Fire restoration options in lodgepole pine ecosystems. *Proceedings—Tall Timbers Fire Ecology Conference* 20:285–297. Tall Timbers Research Station, Tallahassee, Florida.

About the Authors

Stephen F. Arno is a forest ecologist who retired from the USDA Forest Service's Rocky Mountain Research Station after 31 years in federal service. He has studied the effects of fire and the use of prescribed fire and fuel reduction treatments and has authored over 100 scientific publications. He holds a Ph.D. in forestry and plant science from the University of Montana and has practiced restoration forestry on his family's ponderosa pine forest for nearly 30 years.

Steven Allison-Bunnell is a science writer and educational multimedia producer who specializes in natural history and environmental history. He has also written and produced science news for television, radio, and print. He holds a Ph.D. in science and technology studies from Cornell University and a B.A. in biology from the Clark Honors College at the University of Oregon. He is a faculty affiliate of The University of Montana-Missoula Environmental Studies program. He lives with his wife, son, two cats, and sundry houseplants in Missoula, Montana.

Index

DATE DUE
